SCOTT W. STARRATT

Seismosaurus

The Earth Shaker

David D. Gillette

With Illustrations by Mark Hallett

Seismosaurus

The

Earth Shaker

 COLUMBIA UNIVERSITY PRESS   NEW YORK

Columbia University Press

New York   Chichester, West Sussex

Copyright © 1994 Columbia University Press

Illustrations © 1993 Mark Hallett; uncredited photographs
© Southwest Paleontology Foundation, Inc.

All rights reserved

Pen and ink work executed by Dana Geraths.

Library of Congress Cataloging-in-Publication Data

Gillette, David D.

    Seismosaurus : the earth shaker / David D. Gillette ;
with illustrations by Mark Hallett.

      p. cm.

    Includes bibliographical references and index.

    ISBN 0–231–07874–9

    1. Seismosaurus.   I. Hallett, Mark, 1947– . II. Title.

QE862.S3G56   1994

567.9′7 — dc20   93–40318

                                                  CIP

Printed in Hong Kong

c   10 9 8 7 6 5 4 3 2

Contents

Preface and Acknowledgments

One hundred fifty million years ago North America was home to magnificent dinosaurs, winged pterosaurs, and the swimming and paddling reptiles of the midcontinental sea. Now, when camping alone in the deserts of the American West after long hours of excavation, I can almost hear the grunts and groans of these ghosts of the Jurassic. I see a herd of Diplodocus jostling for position at a watering hole or sweeping across a barren landscape in quest of food along the river just over the horizon.

Their bones are all that remain. My hands gently expose them one by one. Each is a reminder of the history of our continent—a world so remote and strange that even poets can scarcely portray the Age of Dinosaurs. But with each bone, our link with this past becomes firmer and the era all the more spectacular.

One hundred fifty million years ago a dinosaur died in what is now central New Mexico. Attacked by scavengers and decomposers, the remains were finally buried beneath the sands of a capricious river. That particular dinosaur now lives again in a new way: it has a name, and it is the subject of an exciting episode in the realm of paleontology.

In the pages that follow, I hope to convey the rich story of how *Seismosaurus hallorum*—one of the biggest of all dinosaurs yet discovered—was reborn as the cherished fossil "Sam." The paleontological portion of the tale begins on a hot wind-swept mesa in 1979. It is a story more of people than bones, more of ideas than facts. And it is an adventure that has dominated my life for nearly a decade.

To those people who have shared in the toil, frustrations, and joys of discovery I am deeply indebted, for they have contributed in many ways to what has come to be called the Seismosaurus Project. Wilson and Peggy Bechtel, working on grants administered by the Southwest Paleontology Foundation, have spent more time on site than anyone else, myself included; since 1988 they have assumed responsibility for the day-to-day management of the excavation, supervision of volunteers, and coordination with government agencies. But for their dedication and persistence this project would have succumbed to financial and organizational difficulties long ago.

All of us involved in the project are indebted to the original discoverers, Arthur Loy and Jan Cummings, who protected the exposed bones and who, with Frank Walker and Bill Norlander, brought them to our attention in 1985.

Among the dozens of volunteers who assisted with the excavation, Kirk Bentson, Bob Webster, and Charles Knapp deserve special recognition for hundreds of hours of service. Additional volunteers include Bertrand and Beatrice Block, Lynett Gillette, Howie Green, Charles Harris, Pam Kuster, Stan Lundy, Joe McDowell, Barry Moore, Matt Mygatt, Lyle Newman, Carol Orr, Hilde Schwartz, Ed Springer, Bob and Linda Strong, Frank Walsh, Woody Weld, Mahlon Wilson, and Gary Yohler.

From Los Alamos National Laboratory (Los Alamos, New Mexico), Roland Hagan and George Matlack were the principal organizers for remote sensing experiments on site and for research in the chemistry of fossil bone. To them I owe a great deal, for they have guided me into a world of technology that is a paleontologist's dream. Many other scientists from Los Alamos contributed in numerous ways. Among them are Harold Bowen, Lawrence Gurley, Rod Hardy, Bill Johnson, Kim Manly, David Mann, Randy Mynard, Carrie Neeper, Don Neeper, John Phillips, Dale Spall, Don VanEtten, and Phil Vergamini.

Two scientists from Sandia National Laboratory (Albuquerque, New Mexico), Cliff Kinnebrew and Wayne Cooke, repeatedly visited the quarry to test their ground-penetrating radar. A team from Oak Ridge National Laboratory (Oak Ridge, Tennessee) consisting of Alan Witten, Joe Sypniewski, and Chris King (dubbed the Seismosaurus Tour) visited the site several times and made important contributions in applications of acoustic diffraction tomography.

Several individuals employed by the U.S. Bureau of Land Management in New Mexico — Al Abbey, Angie Burger, Harry DeLong, Pat Hester, Mike O'Neill, and Dennis Umshler — deftly managed our permits and kept the project within the limits required by law. Especially during early stages of the project, taphonomist Hilde Schwartz contributed to understanding the geological setting. John McIntosh, James H. Madsen, Jr., and Wann Langston helped with identification and advice on sauropod dinosaurs.

Institutional support came from many places: the New Mexico Museum of Natural History, the Division of State History of the State of Utah, Brigham Young University, Los Alamos National Laboratory, Sandia National Laboratory, Oak Ridge National Laboratory, the Smithsonian Institution, the Ghost Ranch Conference Center, the U.S. Bureau of Land Management, and the New Mexico Friends of Paleontology.

Major funding for the project came from the Smithsonian Institution, the National Geographic Society, the National Science Foundation, and the Martin Marietta Foundation. Additional financial support and in-kind contributions from Ann and Gordon Getty, Bertrand and Beatrice Block, Gary Yohler and Pam Custer, Randolph Township Schools (New Jersey), Kirk Bentson, Chuck Harris, Adrian Madera, Frank Beall, and the BTM Wrecker Service and smaller donations from many more contributed to the success of the project.

For friendship and counsel, I am indebted to Wilson Bechtel, Peggy Bechtel, David Thomas, Edwin H. Colbert, Margaret Colbert, Jennifer Gillette, Lynett Gillette, my parents — Dean and Julia Gillette — and Janet Whitmore. Ed Lugenbeel, Laura Wood, and Anne McCoy at Columbia University Press guided me through the process of publication. Freelance manuscript editor Connie Barlow expertly turned my drafts into consistent and manageable text, and to her I am deeply indebted. I thank David Weishampel and Edwin H. Colbert for critical reviews, which greatly improved the text.

Working with dinosaur illustrator par excellence Mark Hallett has been a great pleasure, for he has the uncanny ability to turn words into images that are real and dramatic. All drawings and paintings in this book are Hallett's. Except as noted, all photographs are mine, copyrighted by the Southwest Paleontology Foundation, Inc.

Finally I am grateful to Lynett Gillette and Jennifer Gillette for their encouragement and unfailing support, and to my friend Paul Bowles, now age eleven, whose theories on dinosaur paleobiology challenged me in more ways than he may ever know.

I dedicate this book to all those who have helped in the Seismosaurus Project, and in particular to Paul Bowles and Wilson and Peggy Bechtel.

David D. Gillette

Paul Bowles in 1990. This photograph was taken when Paul was eight years old, during his family's visit to the Seismosaurus site from their home in California.

Wilson Bechtel in 1988.

Peggy Bechtel in 1992. Courtesy of Gary Yohler and Pam Custer.

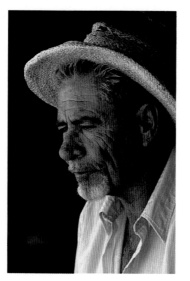

Seismosaurus

The Earth Shaker

I was diplomatic. "There aren't any Morrison dinosaurs known in the area," I told him, "except one partial skeleton excavated several years ago. Otherwise, the closest Morrison dinosaurs are mostly just fragments from eastern New Mexico, two hundred miles away."

Frank was insistent. He offered to show me the snapshots, and if I was interested he would guide me to the site. Next day Frank brought his photos to the museum, and I was immediately convinced. Clearly visible were several vertebrae laid out end to end in a stark white sandstone. They were large, even for dinosaur bones, but from the photos I couldn't tell which part of the body they were from or identify which dinosaur they belong to.

Frank then told me that a television crew for a local documentary news program had filmed the site and that the footage had aired two weeks ago. The public had thus seen the bones in all their splendid detail. He was worried that unscrupulous collectors might now be planning a raid; hence his call. I, too, was alarmed. Frank agreed to take me to the site the next day.

Panorama photograph of the Ojito Wilderness Study Area (from the south). The fossil site is in the low mesa capped by juniper and piñon trees in the midground.

As we drove northwest out of Albuquerque and past the Zia Indian Reservation, towering cumulus clouds gathered to the north over the Jemez Mountains. In these mountains a giant caldera more than twenty miles across is all that remains of what had been a magnificent volcano hundreds of thousands of years ago. The churning clouds now threatened to deliver a payload of thunder, lightning, and flashfloods.

We left the highway and turned west onto a well-maintained gravel road. Within a mile we passed small hills that displayed the awesome power of the earth's crustal motions. What was now the Rio Grande region had once been a major rift and fault zone, as dramatic and powerful in its past volcanic activity and earth movement as is the Great Rift Valley of eastern Africa today. Layers of rocks had been locally folded, contorted, up-ended, and then eroded into a panorama of rich and banded colors: tan, deep purple, brown, white, and red, all incised with gullies and canyons. The line of the fault could be traced to the horizon and beyond for more than a hundred miles. It marked the boundary between two of the major physiographic provinces of North America. We were leaving the Rio Grande Rift and entering the southeastern corner of the Colorado Plateau.

Further down the road the landscape opened into a maze of buttes and broader flat-topped mesas. Some of the mesas were gouged by canyons lined with the deep green hues of juniper shrubs and piñon pines — a welcome contrast to the earth tones of the surrounding desert. I began to identify the geological strata for Frank. The prominent brown layer capping the highest mesas (giving them their characteristic table-top form) is the Dakota Sandstone. Beneath the Dakota lies the softer Mancos Shale, its gray mudstones eroding in gentle slopes too unstable to hold vegetation. Elsewhere in New Mexico these two formations of the Cretaceous Period yield shark teeth and dinosaur tracks a hundred million years old.

The road twisted through a deep canyon, with its own smaller side canyons carved by flash floods. These arroyos had cut through the Mancos Shale and down into the darker Morrison Formation, the youngest of the various Jurassic formations in North America. Its mostly purple and brown sandstones, siltstones, and mudstones were dotted with spring desert flowers. I was pleased: Frank had the geology correctly identified, thanks to Jan. These formations extended across northern New Mexico west into Arizona, east into Texas, and north as far as Utah and Montana. Elsewhere these rocks, especially the Morrison Formation, had yielded a bounty of dinosaurs, but not in New Mexico — at least not yet. I dared to imagine that Frank was taking me to a new bonanza in the Morrison.

Frank turned his pickup off the gravel road and followed the faint tracks of a jeep trail, the same track that Jan and Arthur had used six years earlier. Rain would be welcome, but I hoped the storm would wait till we got back to the highway. Sandy portions of the track showed evidence of recent traffic. A mile farther, after jostling over the rocks and ledges, we parked the truck. There we found more tracks in the sand and the mix of trash one sadly finds just about anywhere vehicles can go.

A closer view of the Seismosaurus site. The resistant sandstone of the cliff face (which was about thirty feet west of the skeleton) is in the upper part of the Morrison Formation, technically called the Brushy Basin Member.

We stopped a hundred yards short of the end of the mesa, on the edge of a barren sandstone ledge. Frank motioned to get out of the truck. I scanned the horizon to line up our mesa with others in the distance. Some high mesas were capped by Dakota sandstone; others, like the low mesa where we had stopped, were capped by resistant sandstones of the underlying Morrison Formation. We were clearly in the Morrison Formation.

Frank led me a few feet below the top of the mesa to a ledge. He carefully removed some small rocks from a pile on the sandstone, then brushed away the loose sand that he and his friends had used, once again, to camouflage the bones after the film crew had finished. In a few minutes he exposed the line of bones. I was transfixed, not because of their size—that realization would come later—but because of how remarkably well they were preserved and because they were still connected.

The vertebrae were uncrushed. Dinosaur bones are almost always distorted from compression caused by eons of burial. We

Close-up view of the ledge just before excavation. The rod in the center of the photograph is two meters tall; the bones lay entombed in the sandstone to the right.

Discovery

counted at least six separate tail vertebrae, lying on their sides. The spine on the top of each vertebra was exposed; the spines were nearly vertical rather than tilted rearward, so I couldn't tell for sure which end of this line of vertebrae was toward the front and which end was toward the rear.

The initial identification had been correct: a sauropod dinosaur, akin to Apatosaurus and Diplodocus. The huge centrum for each vertebra and the tall and erect dorsal spines of great size proved the identification. I estimated that each vertebra was about a foot long. But I couldn't see enough to conclude anything more specific. The only sauropod known from the Morrison in this part of New Mexico was the much smaller Camarasaurus, and since I was working with dinosaur sculptor and museum technician Dave Thomas mounting a Camarasaurus skeleton for exhibit I had no doubt that the Ojito dinosaur was different and therefore new to New Mexico. But I did not expect it to be new to science.

The tail vertebrae as originally discovered. From this perspective the bottom surfaces of four articulated vertebrae, each about twelve inches long, can be discerned. (The carcass came to rest on its side before burial.) Photograph courtesy of Frank Walker.

Closer view of the bottom surfaces of two of the original tail vertebrae. Photograph courtesy of Frank Walker.

Tail vertebrae. The mound in front of the woman contained several heavily weathered vertebrae that had slipped from their original positions before burial. Photograph courtesy of Frank Walker.

I knew we had to excavate these bones—immediately. But there were problems: this was a wilderness study area. Ten thousand acres of the Ojito region were being studied for potential designation as a federally protected wilderness, and during this time strict rules to prevent its degradation were in force. The Bureau of Land Management typically took a year or more to process an excavation permit request for even unrestricted lands; who could predict how long they would take for a wilderness study area? I could imagine rock hounds scouring the area for souvenirs, tipped off by the television publicity, and hauling away in their knapsacks what should be studied and displayed at the new museum — as we dutifully waited for our permit.

At the BLM office in Albuquerque I asked for immediate approval of our application so that a crew from the museum could excavate on Father's Day weekend, only three weeks away. I, of course, told them all the reasons why expeditious treatment was warranted in this case. But I knew my request was unrealistic: to my knowledge no BLM permit for fossil collecting had ever been issued in such a short time. The application had to get clearance from geologists, archaeologists, biologists charged with protecting rare and endangered species, reclamation specialists, recreation specialists, and outside wilderness advisers.

Two weeks later a BLM official called with the news that we

Location map. The Ojito Wilderness Study Area is in northwestern New Mexico about sixty miles northwest of Albuquerque.

could proceed with the excavation as requested. I was as thrilled as I was surprised. We were faced with a few restrictions relating to access and use of equipment, but these demands were reasonable. Immediately, I began to organize a field crew.

It was not, however, the best of times to do so. Construction of exhibits at the museum had reached a frantic pace, and all other concerns were set aside. Curators like me, whose responsibilities included research and collection management, were ordered to wield hammer and saw, or broom and bucket, and join in the madness. At the same time I was managing about fifty outside contracts, all related in some way to paleontology exhibits. We had to schedule the emergency excavation for a long weekend, on off-duty time. Father's Day it would have to be.

My wife, Lynett Gillette (who is also a paleontologist specializing in dinosaurs), and Jennifer Gillette, our thirteen-year-old daughter, happily went along with the idea of a dinosaur excavation for our Father's Day plans. Museum sculptor and paleontological technician Dave Thomas celebrated Father's Day on site too. Several BLM employees chose to join us, including recreation specialist Angie Burger and Dennis Umshler, a geologist who would later guide us through additional permit applications. Dennis agreed to be photographer for the weekend. Rounding out the crew was an assortment of volunteers from the fledgling and informally organized New Mexico Friends of Paleontology. The original discovery team—Arthur, Jan, Frank, and Bill—were also honored guests and helpers.

The size of the crew matched the size of the job, which we were going to do without dynamite. In my experience excavation of bones this size entailed considerable labor and no small amount of seat-of-the-pants engineering. But I saw no reason to consider dynamite—a tactic that would have set off alarm bells for those processing the permit. My plan was to use small jackhammers powered by generators to expose blocks of rock around the bones. The blocks would then be wrapped in plaster-soaked burlap for transport. Pickup trucks would be sufficient for hauling everything in and out, and we needed little else except lots of drinking water, brimmed hats, and sunscreen.

Initial excavation in 1985. This drawing is based on the only surviving photograph.

We arrived early on Saturday morning. Each participant had one or more specific jobs: digging, operating a jackhammer, mapping and note taking, cutting burlap sacks into strips, preparing plaster, taking photographs. In the heat and surrounded by gnats, we set to work.

The sandstone was hard around the bones, softer a few inches away. This made the digging easier than we expected. By midmorning we had outlined eight vertebrae, following the bones deep into the rock to determine their limits. Next we selected where to make the separations in the string of vertebrae and how to undercut the blocks for plastering. Where possible we used natural cracks in the rock — sometimes opting to break the blocks by way of a natural fracture running right through a bone rather than attempting to force a break (and thus risk additional fracturing) between articulated bones. Repairs could be made in the laboratory, and the breaks matched with precision.

By midafternoon Sunday we had isolated all the exposed bones. We then wrapped them in plaster and burlap. Several blocks weighed as much as four hundred pounds. After securing the blocks in plaster and burlap, we loaded them onto a trailer and pickup trucks, using four or five strong people for each block. Tires bulging and springs overstressed, our trucks and the trailer made the slow trip back to Albuquerque as the sun set.

Two long days of excavation proved to be just the right amount of time for removal of the exposed bones. The excavation revealed that these bones came from the middle part of the tail, but their exact position could not be established until laboratory preparation — a tedious and time-consuming process — was completed. Little did I realize that those two days were to lead to months and years of work at the site, excavating more bones in collaboration with dozens of other research scientists.

I suspected that the line of the tail would continue into the mesa and that it might, if we were very lucky, lead to the rest of the body. One partly exposed bone fragment at one end of the original line of vertebrae suggested that the eight bones we removed that weekend were not the only bones there. Only a few square inches of that mystery bone were exposed, and I didn't want to jeopardize its security in the rock by exposing the bone further without doing a job sufficient for collecting it. By Sunday afternoon we realized we couldn't expect to collect any more bone on that initial outing and that expanding the quarry was not practical.

That bone fragment became an impetus and a marker for future excavations. Along with the fact that the collected vertebrae were perfectly articulated, this fragment was tangible evidence that more bone might be present. On it hung our best chance for securing the kind of financial assistance that a full excavation would demand.

We stored the plaster jackets containing the bones in the garage of one of the volunteers, as I was uncertain how to proceed. I was overloaded already with my share of exhibit and construction management at the museum and could not even consider beginning careful work on the bones in the paleontology laboratory. Nor would space be available to store the massive blocks in the new building.

Dennis Umshler called a few weeks later with disappointing news. None of his photographs were properly exposed. Our official photographic record of the initial excavation had been lost to the dual distractions of gnats and heat.

"Don't worry," I reassured him. "This was just a routine excavation, and I understand that some of the other volunteers got a few snapshots."

The excavation had been conducted under my guidance, but it had required the gift of many people's labor. The matter of identification was now my responsibility alone. The identity of this dinosaur was, however, still a puzzle to me. At the time I suspected that the specimen's prime value was not its identity but the facts that its bones were articulated and it was new to New Mexico. Dinosaur bones are usually scattered and disconnected, owing to the work of scavengers or to stream action. Only rarely do we find bones connected, and even then the skeletons are seldom complete.

Even specialists cannot always identify a fossil with confidence at the time of discovery. This was my first experience with a sauropod dinosaur skeleton. Paleontologists usually try to make a reasonable guess as to the level of family or genus, and we are often correct, but unless the diagnostic traits are evident on site, firm identification must await laboratory analysis. In the case of an organism new to science, our usual first reaction is puzzlement, as we try to relate the observed characteristics to what we know about established genera or species. The recognition of a new taxon almost never occurs at the time of discovery. This sauropod dinosaur clearly belonged to the family Diplodocidae, but whether it was Diplodocus, Apatosaurus, Barosaurus, or one of the other lesser-known genera or a new genus altogether was unclear. I could resolve this problem only by comparing these bones to sauropod bones in other collections.

In the meantime this partial, but articulated skeleton would become the focal point for more discussion than any other dinosaur I had ever excavated. Before taking my position at the new museum in 1983, I had spent twenty years specializing in Ice Age animals, Cretaceous sharks and fish, and the natural history of vertebrates. In this new post I expanded my research goals to include dinosaurs. By the time of our excavation I had studied thousands of dinosaur tracks, including those of sauropods, and I had excavated a partial skull of *Tyrannosaurus rex* and perhaps a dozen other dinosaur skeletons and partial skeletons. Since then I have worked in all three periods of the Mesozoic (Triassic, Jurassic, and Cretaceous) on hundreds of dinosaur skeletons, mainly in Utah and New Mexico. But this excavation has been the most important.

Names are essential for communication. Unable yet to give these bones a scientific name, I needed at least a familiar name. I could have referred to the skeleton by catalog number NMMNH 3690 (New Mexico Museum of Natural History paleontology specimen number 3690), or by a field number that would correspond to notes taken during the excavation. But like paleoanthropologists working on hominids in Africa, I felt the need for something more personal, more familiar. Lucy got her name from a famous Beatles song popular at the time of discovery; Twiggy was an obvious name for a hominid with a flattened cranium found at a time when the fashion industry's most famous model was so named for her twiglike, flattened physique. Even something purely geographic (for example, the Taung baby) would suffice.

My choice for the Ojito dinosaur was Sam. A common nickname for Samantha or Samuel, it was appropriately ambivalent for an individual dinosaur that was, as yet, known from just eight bones. But soon it would be time to give Sam a proper scientific name.

Chapter *2*

Naming the New Genus

In 1986, nearly a year after our Father's Day excavation, I was awarded a grant from the Smithsonian Institution. Its "short-term visitor" program encourages scientists to study the Smithsonian collections. At the Smithsonian I studied the sauropod dinosaurs and accumulated critical measurements for comparison of Diplodocus and other dinosaurs with the bones of Sam. I extended the travel on that grant to include the American Museum of Natural History (New York City) and the Carnegie Museum of Natural History (Pittsburgh), where the most important specimens of the sauropods Diplodocus, Apatosaurus, and Barosaurus are on display and in study collections.

The information I gathered on this extended travel proved instrumental in my determining the uniqueness of the new bones from New Mexico, and it gave me a firm basis for making size comparisons. According to all comparisons I could make with Diplodocus and Apatosaurus, Sam would be considerably longer, and probably larger in all dimensions. I thus knew we faced a colossal excavation if the bone fragment indeed marked the position of the tail and would lead to the remainder of the skeleton.

Among dinosaurs, sauropods were both the largest and the most conservative. With long necks, long tails, capacious rib cages, and legs like pillars, sauropods changed little physically from the time of their first appearance more than two hundred million years ago (early Jurassic) until their extinction, along with all the other dinosaurs, sixty-five million years ago (end of Cretaceous). They were always enormous and they were all plant eaters. A 1990 summary by J. S. McIntosh lists most of the currently accepted genera of sauropods, along with their ages, principal localities, and comments concerning distinguishing features. (Seismosaurus was not included in that list because its formal description had not then been published.)

Sauropods like Sam flourished in the Jurassic and reached their zenith in diversity at the end of that period, roughly one hundred fifty million years ago. Sam was one of many kinds of sauropods that lived in the late Jurassic. Skeletons of late Jurassic sauropods occur in (relative) profusion in western North America, principally in the Morrison Formation, and in eastern Africa in the Tendagaru beds of Tanzania. The Jurassic Period could well be called the Age of Sauropods.

Sauropods never recovered from a decline that began in the early Cretaceous when angiosperms (flowering plants) began to replace the conifers, cycads, and ferns that had dominated earlier landscapes. Perhaps sauropods survived in the ever-dwindling archaic habitats, while other herbivores with adaptations better suited to feeding on angiosperms replaced them as the dominant plant-eaters of the Cretaceous. Their ultimate disappearance at the end of the Cretaceous scarcely affected the course of history, for their abundance and diversity had diminished so severely that their scant contribution to the dinosaur world at the time of the great extinction could be easily overlooked.

More than 90 genera and 150 species of sauropods have been named, but only a few are relatively well known scientifically. Typical among the Jurassic sauropods was the genus *Camarasaurus*, a stout animal with relatively short neck and tail. Apatosaurus (as *Brontosaurus* has been renamed) and Diplodocus belong in another family, differing in their exceptionally long necks and tails and more delicate skulls. All these giants lived in the western United States, likely roaming in herds and feeding constantly on conifers, cycads, and ferns. The weight of most sauropods typically approached twenty tons, with the heavily built Apatosaurus weighing considerably more. Barosaurus, a close relative of Apatosaurus and Diplodocus, had an exceedingly long tail, sometimes described as a whiplash.

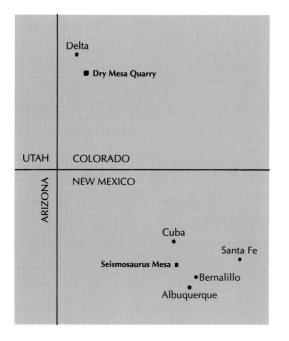

The type locality of Seismosaurus in the Ojito Wilderness Study Area northwest of Albuquerque, New Mexico, and the Dry Mesa Quarry southeast of Delta, Colorado—the type locality of Ultrasaurus macintoshi, Supersaurus viviani, and Dystylosaurus edwini.

The giants were common, the supergiants rare. Brachiosaurus is the best known of the supergiants. It weighed about twice as much as Apatosaurus — about seven to eleven times that of an elephant. With long neck and forelegs, it had a more giraffelike build than the other sauropods. Also among the supergiants were Ultrasaurus, a close relative of Brachiosaurus, and Supersaurus, more closely related to Diplodocus. Ultrasaurus and Supersaurus are both known only from Dry Mesa Quarry in Colorado. A third Dry Mesa supergiant, Dystylosaurus, rivaled the others in size, but like Supersaurus and Ultrasaurus it is known only from isolated bones. Seismosaurus is the fifth supergiant sauropod — and the only one besides Brachiosaurus known from partial skeletons.

Skeletons of the giants (Camarasaurus, Apatosaurus, Diplodocus, Barosaurus) suitable for display in the great museums were among the trophies sought by collectors during the rush for dinosaurs in the American West during the late 1800s and early 1900s. Casts of Diplodocus were traded to museums around the world, making it one of the most well-known dinosaurs. Today, visitors can see partially articulated skeletons of these and other dinosaurs on display in situ, locked into the sandstone from the Jurassic river bed in which they were buried.

This spectacular demonstration of dinosaur bones at Dinosaur National Monument, on the border of Utah and Colorado, documents one of the most productive sites for Jurassic dinosaurs in the world.

Mounted skeletons of sauropods now grace the halls of museums around the world. Increasingly, these skeletons are replicas made from original reconstructions. And original reconstructions themselves are not the real thing; they are most often a potpourri, a single skeleton made from isolated bones — often from several individuals. Reconstructions of skeletons are not the same as restorations. These two terms have very different meanings. For example, John Harris distinguishes them in this way: "the term *reconstruction* is used in the sense of piecing together the original but often fragmentary fossilized parts of extinct animals, whereas *restoration* is used to describe the depiction of their original appearance — muscles, flesh, skin, and all." Thus, a reconstruction is a skeleton, and a restoration is the animal in the flesh, as though living. Animated restorations, which depict dinosaurs in their flesh-and-blood glory with movements controlled by computers, draw huge crowds. Artists' paintings, meanwhile, have long re-created the living animals in their habitats — giving them a lifelike quality that is sometimes

Skeletal reconstruction of Diplodocus carnegii*, probably the closest known relative of* Seismosaurus. *This slender sauropod is known from essentially complete skeletons from the Morrison Formation. Its skeletal anatomy is well established, and it formed the basis for most comparisons with Sam's skeleton.*

so realistic that they seem to be photographs taken by a camera loaded with fast, color-saturated film.

But the real nuts-and-bolts of the dinosaur world is skeletons and the sites from which the skeletons have been recovered. The real natural history of dinosaurs resides there, in the bones and the collecting localities, the only sources for the raw material of dinosaur studies. The Smithsonian grant gave me the opportunity to study the real bones of some of Sam's relatives. This award was pivotal in the project, and it led to the other, larger grants that made excavation beyond the discovery stage possible.

Before I went to the Smithsonian, I thought Sam was a member of the genus Diplodocus, a well-known and widespread dinosaur in the Morrison Formation of Utah, Colorado, and Wyoming. But I had trouble matching the eight tail vertebrae we had collected during our initial excavation to known tails of Diplodocus. The proportions were off, the dimensions were too large, and the anatomical details were different enough to question the identification of Sam as Diplodocus. The possibility surfaced: maybe it's a new dinosaur. The fact that the site is hundreds of miles from other Diplodocus localities in the Morrison Formation added to the uncertainty.

Eventually I came to believe that Sam could not be Diplodocus or kin such as Barosaurus or Apatosaurus. The tail bones of Apatosaurus do not have a deep concavity on their undersurface. The tail bones of Barosaurus have the deep concavity, like Sam's, but the vertebrae are relatively short. The closest resemblance was to Diplodocus, but the differences were still too great. Notably, Sam's vertebrae are proportionally longer and taller, and the dorsal spines are nearly erect, quite in contrast to the tail vertebrae of Diplodocus. When I consulted with other sauropod specialists, they were unable to offer any new information or interpretations that I had not already considered. I thus concluded that Sam belonged to a hitherto unrecognized species of dinosaur.

When the New Mexico Museum of Natural History decided to announce the existence of this new and impressive dinosaur to the public later in 1986, I faced a dilemma. We needed a name for the skeleton, and I could not assign it a name based on any known dinosaur.

The paleontologist who first describes a fossil as a new species has the singular responsibility and honor of selecting the name. By international convention, the name should be latin-

Muscular anatomy of Diplodocus carnegii. *This kind of drawing is a prerequisite for producing a reasonable in-the-flesh restoration. Seismosaurus was similar, but had more massive hips, stouter (but not longer) legs, and a tail that differed in some important ways.*

Naming the New Genus

ized according to a universal standard adopted by all zoologists, *The International Code of Zoological Nomenclature.* Its contents read like a legal document, and issues related to naming of animals resemble court cases.

The choice of name — the technical name — for a dinosaur, or any fossil organism, is as important to a paleontologist as naming a newborn is to a parent. We cannot take the matter lightly, because the technical name will stay with the species forever. Sam and all Sam's kind would be known by the technical name. The name should have meaning; ideally it should also be easy to remember and pronounce. And it should be constructed using the agreed-upon rules designed just for this purpose. In some ways the honor of coining the name is the most pleasant of our responsibilities.

That summer, before the press conference organized by our museum staff, I lay awake at night in my cabin in the mountains east of Albuquerque, deliberating. I pored over my dictionary of scientific names, seeking an appropriate root to combine with *-saurus* in keeping with the tradition of Apatosaurus, Barosaurus, Camarasaurus, and dozens of other dinosaurs named under this convention.

After several weeks of searching for a name, I made a decision. I chose Seismosaurus. *Seismo* is the Latin root for "shaking." It is familiar in words like *seismic* and *seismology*, all relating to ground-shaking generated by earthquakes or underground blasts. Sam would be the earth-shaking dinosaur. I searched the technical literature to ensure that this name had not already been taken; if so, the name would have been "preoccupied" and not available for any newly discovered species.

I was lucky. Seismosaurus hadn't been used before; my first choice was available. As long as I didn't attach a species name to this informal genus, any publication of the name in print would be safe from technical nullification — provided I was right that Sam was a new genus. I would use this name in the press conference.

Restoration of the in-the-flesh anatomy of Diplodocus carnegii. *(Color patterns are conjectural.) In life,* Diplodocus *would have been hard to distinguish from Seismosaurus except that the largest adults among the latter were probably 20 to 50 percent longer, with a disproportionately long neck and tail and stouter legs.*

Floodlights blinded me, and the array of microphones seemed like menacing tentacles of a giant octopus. Paleontologists are not trained in graduate school for press conferences. I briefly explained what we had found, unveiled the bones, and revealed the new informal name. I reported that this individual of Seismosaurus was probably at least 110 feet long, comparing it with our own mounted skeleton of Camarasaurus (which is only

about 50 feet long.) This put Sam in the ranks of the super-giants, not the giants.

I then demonstrated how I made that calculation by direct proportions with Sam's closest relative, Diplodocus. Each vertebra was at least 20 percent longer than corresponding vertebrae in Diplodocus, and with the disproportionately tall neural spines, there was the possibility that the overall length of the tail (and indeed the entire body) was also disproportionately long. Although I had data in hand that indicated a more likely length of 120 feet or more, I chose to conservatively estimate Sam's length at 110 feet—which is longer by 23 feet than the longest specimen of Diplodocus, the previously accepted longest dinosaur. (Later I would revise these figures upward.)

Questions came from all directions as each reporter sought a different angle to develop. Sam's new name, Seismosaurus, caught on immediately, capturing the reporters' imaginations. One reporter asked rhetorically why I didn't select a name like Superdoopersaurus to follow the recently invented names Supersaurus and Ultrasaurus for two supergiant dinosaurs that were discovered in Colorado. That quip lightened up the discussion.

We had prepared an exhibit case for the four vertebrae, set beneath the Camarasaurus skeleton for comparison with its tail

First display of the tail vertebrae at the New Mexico Museum of Natural History, Albuquerque. This vertebra is no. 20, counting from the base of the tail. It was the anterior-most of the original eight excavated in 1985.

vertebrae. Comparison with a skeleton of Diplodocus or Apatosaurus would have been more appropriate, since Seismosaurus belongs in the same family with these two familiar dinosaurs, but the Camarasaurus skeleton was the closest comparison we could make with the exhibits available. The corresponding vertebrae in the tail of the Camarasaurus skeleton are ridiculously small by comparison. The display emphasized the extraordinary size of the new dinosaur, even though we had only four bones to present to the public.

The press conference generated a surge of media attention, more than I ever imagined. Sam (rather, Seismosaurus) was spectacular.

My allusion to "earth-shaking" proved ironic, for the next year we would initiate experiments in remote sensing to look for more of Sam's bones hidden deeper in the mesa. Artificially generated sound waves (from a fancy shotgun) would help us "see" bones without digging. That technology was called seismic tomography.

To formalize the name Seismosaurus I needed to fully describe the eight tail vertebrae on which the determination had been made. This description had to be published in a technical journal. I had to distinguish the bones from all other dinosaurs, including Sam's closest relatives, Diplodocus, Apatosaurus, and Barosaurus. This might appear to be a dry and simple task, but technical descriptions of new species are difficult and demanding. Putting into words the description of a bone or skeleton is an extraordinary challenge of communication in the use of our wonderfully versatile language. To succinctly describe an object that is as irregular as a bone, and to do so in words that others can understand without ever seeing it, is immensely satisfying. I, of course, had a science illustrator draw the bones from various perspectives, but in a scientific journal the words are definitive.

Formal description would not be done overnight. What is more, I knew that if I waited I would almost surely have more bones upon which to base the genus-making description. Publication would therefore wait.

Seismosaurus was not, however, a complete name for formal publication. Living and extinct organisms are given binomial names. The principle of binomial nomenclature is that every organism is given a pair of names: a genus name that is capitalized and a species name that is not. Humans belong to the genus Homo. We share this genus with no living species but with several extinct hominid species — *Homo erectus*, for example. Our species is sapiens. We are therefore *Homo sapiens* — presumably, "wise" humans.

The informal name Seismosaurus established at the press

conference could not be formalized without a species name to follow it. That left me with another decision: what to call Sam's species in the formal description.

Sometimes a species name is coined for an anatomical feature, or for a locality, or for a person, such as the discoverer or a patron of the project. Several times I half-jokingly offered to name the species after anyone willing to donate half a million dollars to the project, and the genus after anyone willing to donate a million dollars. My circle of friends is not wealthy; I got no takers.

I considered naming the species for Arthur Loy and Jan Cummings, who together found the bones. In fairness, however, I recognized not only Arthur and Jan as the discoverers but also their friends Frank Walker, who brought the bones to my attention and showed them to me, and Bill Norlander, the fourth member of this fraternity of hiking buddies. I couldn't name the species for all four, and naming it for one wouldn't be fair to the others.

What about geography? The correct latinization of *ojito*, for the site in the Ojito Wilderness Study Area, would be *ojitoensis*. The pronunciation would thus be a puzzle to everyone not familiar with Spanish etymology, and I dislike tongue-twister names anyway.

What about anatomy? Most of Sam's anatomical features are subtle and I couldn't find any one trait in particular that would by itself characterize the species. I thought about referring to the size of the new species by using *longus* or *colossus*, but these names didn't seem appropriate either.

I settled on naming the species for the Reverend James Hall, director of the Ghost Ranch Conference Center, and his wife Ruth Hall, an amateur paleontologist who inspired several professional careers by her teaching. Ghost Ranch is in northern New Mexico, a study center owned by the Presbyterian Church in the canyonlands north of Santa Fe made famous by the artist Georgia O'Keeffe. On its 23,000 acres is one of the richest and most spectacular dinosaur sites in the world, a quarry where at least a thousand individuals of the little predatory dinosaur Coelophysis have been excavated. Jim and Ruth together supported paleontology in and around Ghost Ranch and northern New Mexico for a quarter century. I began working there in 1985 and continue with several active projects related to the Coelophysis quarry. Ghost Ranch has since established the Ruth Hall Museum of Paleontology, organized by Lynett Gillette, the museum's first curator.

Seismosaurus would thus bear the simple species name "halli." *Seismosaurus halli,* or "Hall's earth-shaker dinosaur," it would be.

I now had the name necessary for publication, and soon I had more bones. I submitted my description to the *Journal of Vertebrate Paleontology* in 1989. The review process stretched on, however. Scientists are naturally skeptical of claims of new species (and, even more so, genera), and the peer reviewers of my paper took their task seriously. I had to respond to their criticisms; producing acceptable revisions added another year to the publication date. The descriptive paper was finally published in 1991. Prior to that time, I had given a talk at a scientific symposium (1986) and had published a short abstract (1987), referring to Sam as "a giant sauropod" or "a new giant sauropod" from the Morrison Formation of New Mexico. I had also written a popular article on the excavation for the *Ghost Ranch Journal.* But publication of a formal description and a full scientific name in the *Journal of Vertebrate Paleontology* made it official.

To paleontologists the full and correct name for the new species is now "*Seismosaurus hallorum* Gillette 1991." My initial name proved to have an incorrect Latin ending of a genitive singular—a mistake recognized by George Olshevsky, a dinosaur classification aficionado. So it was changed to the plural form. I formally assigned it to the family Diplodocidae, the family that includes the giants Diplodocus, Barosaurus, and Apatosaurus and the supergiant Supersaurus, all from the Morrison Formation of western North America.

From my original coining of the name Seismosaurus to technical publication of the name *Seismosaurus halli* (more properly, *Seismosaurus hallorum*) took five long years. With the formal publication of the name and the description of Sam's bones as the basis for the new species, our initial goals had been achieved: we had defined the species, identified its distinguishing characteristics, established the geologic age in which the animal lived (late Jurassic) and the geographic position of the site (the southern end of the Morrison Formation), and presented the data and the interpretive calculations that would verify Sam's size — then calculated as between 128 and 170 feet, or between 39 and 52 meters — in the technical literature, through the rigors of peer review.

Sam's position in the scheme of classification of animals can be succinctly summarized. The fundamental unit of classification is the species. Taxonomic categories above the species level are increasingly subjective, generally arranged in hierarchical order. For Sam, the full classification using traditional ranks is as follows:

Phylum Chordata
Subphylum Vertebrata (all animals with backbones)
Class Reptilia (all reptiles including dinosaurs)
Order Saurischia (the giant quadrupedal herbivorous dinosaurs and the bipedal carnivores)
Suborder Sauropoda (the giant quadrupedal, long-necked herbivores)
Family Diplodocidae (relatives of Diplodocus)
Genus *Seismosaurus*
Species *hallorum*, correctly expressed as the binomen *Seismosaurus hallorum*.

Some paleontologists prefer to separate dinosaurs from the reptiles into a distinct class: Dinosauria. Usually, by that convention, Dinosauria includes only dinosaurs, but some paleontologists place birds in the same class, subsuming the traditional class Aves into Dinosauria. At issue are the questions of origins and the philosophy of establishing these evolutionary hierarchies. In recent years applications of the principles of cladistics, and with them a new system of nomenclature, have clarified many questions of ancestry, but the basic unit of classification, the binomial (genus and species) remains largely unaffected.

This naming, this classification, this identification and formal description of Sam depended ultimately on the bones. How does one find more bones hidden, and perhaps scattered, inside a mesa?

Chapter **3**

Help!

In 1985 the curators of the soon-to-open New Mexico Museum of Natural History were each invited to present a short seminar at nearby Los Alamos National Laboratory. Expecting that the scientists attending my presentation would have little interest in the anatomy of Ice Age glyptodonts or the biogeography of Miocene sharks in the Caribbean (two of my research projects to date), I chose instead to discuss local paleontology. More important, I decided to share two problems. These were not paleontological problems in the usual sense, but problems that had bothered me since I began fieldwork as an undergraduate student in 1967.

Usually, when invited to present a seminar, scientists talk about their latest achievements. That is, after all, a good way to inform others about our work. Often, however, our goals are more subtle: we also intend to impress the audience, our peers, with our prowess. This professional exposure is important, and promoting one's work in hope that it will be useful to (and, ultimately, cited by) others is a stimulus to scientific advance. And if our presentations yield additional benefits, such as job offers or a more enthusiastic review of a proposal for funds, so much the better.

To present *problems* at a seminar is unorthodox, because in so doing we reveal our weaknesses. Often in technical seminars, members of the audience take great delight in publicly pointing out deficiencies or inconsistencies in the presenter's research, exposing weak points in methodology or logic. Nevertheless, I did just that: I confessed my lack of knowledge on two peripheral but bothersome subjects, thinking maybe my talk would spark some interest or even lead to ideas for technological applications I had never considered. This was, after all, Los Alamos National Laboratory—the place where the atomic bomb was born, and the place that had continued to bring in brilliant scientists to fight the Cold War.

I stood in the auditorium before a gathering of about a hundred scientists and technicians, hoping that the combination of "big dinosaur," "local excavation," and "looking for technological ideas" would pique their interest. I posed two specific questions. First, is it possible to see traces of soft tissue that may be preserved next to the bones, perhaps as unseen chemical signatures of the outline and position of muscles or stomach?

Occasionally soft parts of vertebrate animals are preserved with bones: fossil tendons are common; wing membranes of pterosaurs and bats have been recognized with some skeletons; feathers have been found with fossil birds. In a few cases, contents of the body cavities have been preserved. As a graduate student I had published a paper that described three Cretaceous fish fossils as containing probable egg masses (in a research project where I puzzled over the chemistry of preservation). And I am still intrigued by the problems of chemical alteration in fossilization.

The idea that ghost outlines might be preserved with a skeleton drew considerable interest at the seminar. I could feel the abrupt change of attention in the audience. No longer were they polite and dutifully courteous. Here was a problem these scientists could relate to, a challenge that might bridge the gap between my Victorian-style approach to fossils and their world of high technology. I sensed their tension, and my confidence grew. I went into more detail than I had planned: I gave an overview of the problems of preservation chemistry. How, in fact, do bones become fossilized?

My allotted time was short, however. I moved on to my second question: Is there any way to "see" into the ground before excavating a skeleton? Is there some technology that can give me a kind of X-ray vision so I can know whether and where to dig?

I told them why I was interested; I told them about the gigantic tail vertebrae that had been discovered an hour's drive of Albuquerque. I told them why the articulated character of the bones made them specially valuable — and seductive. And I told them of my hopes of following the tail forward, into the mesa with its ten-foot cap of sandstone.

I knew it was a ludicrous wish: to see a skeleton beneath the ground before striking the first rock with a pick-axe. With the frustrations of a century of paleontologists before me, I conveyed to my audience the difficulties we faced in excavating this exceptionally large sauropod: a ten-foot wall of sandstone to move, wilderness advocates demanding minimum disturbance and no mechanized equipment, and the likelihood of needing to race to complete the excavation before the area becomes a formally designated wilderness. If we could see the buried bones in the ground before excavating, we could dramatically improve our efficiency and minimize the disturbance.

At the conclusion of my fifteen-minute presentation, I asked for help. I expected one or two takers. Instead, I was swamped with volunteers and ideas. They overwhelmed me with enthusiasm. That seminar proved to be the most productive quarter-hour talk in my career.

Los Alamos scientists took up the challenge. On a field trip to the site, Nate Bower, a contract researcher from Colorado College, found a bone chip that he took back to his lab for chemical analysis — the results would prove surprising. Carrie Neeper, a microbiologist from the city of Los Alamos (but not at that time employed by the lab), became one of the local coordinators for volunteers and information sharing. Later, geologist Kim Manley, also from the town of Los Alamos, took an interest in gastroliths.

News of my talk and my challenging questions went beyond Los Alamos. Roland Hagan, an electronics technician at Los Alamos, enlisted the collaboration of Cliff Kinnebrew and other scientists from Sandia National Laboratory in Albuquerque to join with Los Alamos in their radar experiments. Later, Roland invited scientists led by Alan Witten from Oak Ridge National Laboratory to try their hand with technology still under development for locating buried hazardous wastes and other classified applications. By 1987 the friendly rivalry between the scientists from these three national laboratories seemed to be producing tangible ideas for assisting the excavation of Sam.

The Seismosaurus excavation had become THE SEISMOSAURUS PROJECT, a multifaceted experiment involving not just traditional paleontology, but also chemistry, physics, engineering, electronics, and a little bit of magic — magical science and magical friendships.

On one visit to the site by Los Alamos scientists, chemist Shaun Levy took hold of the fact that dinosaur bones are often preserved with relatively high concentrations of uranium. An earlier analysis at Los Alamos established that Sam's bones contain a small amount of uranium, too. The origin of this uranium is somehow related to percolation of ground water long after burial, but the actual process of deposition and concentration is problematic. Because some uranium-containing minerals fluoresce under ultraviolet light, we wondered whether Sam's bones had adequate concentrations of uranium and the right minerals to fluoresce.

We collected a fist-size fragment of bone on-site, and I accompanied the group back to Los Alamos to witness this test of fluorescence. We needed only an ultraviolet lamp and a place dark enough to conduct the experiment. Someone suggested the men's room. It's only big enough for two people, or uncomfortably, maybe three, but it can be made absolutely dark. So, several of us crowded in, turned on the ultraviolet lamp, and turned off the lights.

The fossil bone glowed. Whether the fluorescence came from the uranium was still uncertain, but at least we had discovered an unusual and potentially important property of Sam's bones, and perhaps many fossil bones.

Our discovery that dinosaur bones can fluoresce, we learned later, had been made by rock hounds long ago. This fact was well known by amateur collectors, a spin-off from the widespread use of ultraviolet lights to prospect for certain minerals in mines and caves. This fluorescence was new to us, however, and we soon learned that the glow comes not from uranium minerals in the fossil bone (uranium is there in significant concentrations, to be sure, but not in minerals that fluoresce), but instead from the natural fluorescence of the hydroxyapatite, a crystalline mineral found in all living bone—and, incidentally, probably all fossil bone in its original or nearly original state.

The discovery of fluorescence in Sam's bones suggested an immediate practical application. Because the bones were buff-colored and difficult to distinguish from surrounding rock, perhaps we could use ultraviolet light to prospect for more bone. So on a dark, moonless night our team of prospectors waited until nearly midnight to try so-called black lights we brought from Los Alamos. In three small groups, armed with flashlights to guide us to the broken cliff face and black lights to search for bones, we spread out over the site. One group searched where bones had already been excavated. Another searched where bone fragments were known to be exposed—and which we had specifically marked for testing that day. The third group searched on the face of the cliff.

The experiment was wildly successful. We found bone everywhere—most of it in small fragments that had weathered out and disintegrated over the past thousand years. Some of the bone we hadn't seen before, but none of the discoveries actually led to new intact bones in the mesa. Nevertheless, I was delighted. The night's work had made me confident that the skeleton had not been exposed and eroded away with boulders and rocks and pebbles in the cascade of rubble on the hundred-foot slope beneath the site. Rather, if more bones did accompany the eight tail vertebrae, then they were still safely preserved within the mesa—albeit beneath a cap of rock that would make life difficult for the excavation crew.

Another spin-off from this discovery of fluorescence helped us improve our laboratory preparation of several of Sam's vertebrae. One vertebra from the tail was encased in rock that was especially hard. Removal of that rock would be difficult; the work would be slow and tedious, progressing by only a few square centimeters a day. The problems were compounded by the intricate folds and projections of the bone, which were tough to follow without damaging the bone's surface. More frustrating, however, was a peculiar condition of preservation that we found on many of the upper surfaces of the bone throughout the skeleton: the sandstone rock actually penetrated the fabric of the bone, through an interval of several millimeters, destroying the naturally sharp contact between bone and rock that is common to most fossil bone. To make matters worse, the rock and the bone were identical in color, and almost identical in texture. We found that a technician could easily dig right through the bone structure and never realize it.

To solve that problem we improvised an experiment using ultraviolet light to see whether we might readily distinguish bone from rock in the laboratory, where the majority of bone extrication must be done. In the makeshift black box, which blocked out all ambient light and allowed only ultraviolet light from an overhead fixture, the bone glowed a brilliant blue and orange tint. The surrounding sandstone remained dark and unreflective. With the aid of the black light in otherwise total darkness, Wilson Bechtel prepared that vertebra with delicate accuracy and efficiency; the rate of exposing the bones improved to as much as a square inch a day, sometimes even more. We were elated.

These modest beginnings eventually led to a major research investigation of the chemistry of fossil bone preservation, including the vexing problem of why uranium accumulates in fossil bone. A preliminary chemical analysis of the fossil bone fragment I had casually given to Nate Bower was surprising: a

A partially prepared Seismosaurus caudal vertebra. The sandstone in which this bone was encased was so perfectly matched in color and texture that distinguishing bone under ordinary lighting conditions (such as photographed here) was almost impossible.

Help!

The same vertebra under low-intensity ultraviolet light and reduced natural lighting.

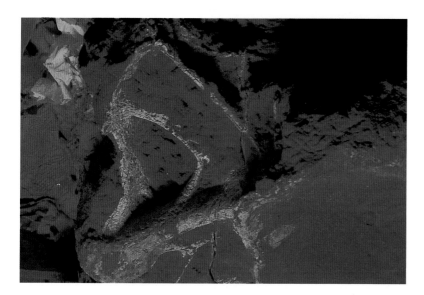

Close-up of ultraviolet image. The brilliant fluorescing purple is bone and the nonfluorescing material is sandstone. Wilson Bechtel completed the meticulous preparation of this vertebra in a makeshift box that was illuminated only by ultraviolet light. This visual enhancement doubled or even tripled his efficiency. Courtesy of Wilson Bechtel.

dozen major elements in the composition of Sam's bones were of the same concentrations as that in samples of modern bone. The match was, in fact, almost identical. The conclusion was inescapable: the dinosaur bones could not have been replaced by secondary minerals. These were not stone bones; these were real bones. A large portion of what remained of Sam must be original material.

Research concerning preservation chemistry began with Nate Bower's report, which stimulated Los Alamos to get further involved. They, in turn, recruited colleagues to take up the challenge I had presented in seminar several months before —

now modified to address the entire question of geological processes that lead to bone preservation. I remember having been taught about the mysterious process of "molecule-by-molecule replacement" believed to occur in fossilization. But when pressed, I could only confess confusion.

From these rather casual beginnings George, Roland, their colleagues, and I began to identify specific problems that required carefully controlled experiments. We recruited other scientists, asked for advice, held rump session seminars, and prodded colleagues well beyond the bounds of New Mexico to lend a hand. But research into chemical preservation of fossils was not the only spin-off of my Los Alamos seminar.

A century ago during the golden age of dinosaur excavations, and even thirty years ago, the principal or even sole objective in dinosaur excavations was the procurement of exhibit specimens. Today, however, the sedimentary context holds equal importance to the bones. Paleontologists now give attention to habitat interpretation (largely for improving our understanding of behavior). And we try to accurately correlate the stratigraphy of a new site with previous excavations (as precise data can be used to understand regional or even global biotic changes such as migration and extinction).

Whereas our objectives are different today, the techniques of excavation have changed little in the past century. We still use hammer and chisel, pickax and shovel — all powered by muscle and cooled by sweat. Sometimes now we do use jackhammers, driven by generators and compressors. Plaster-and-burlap bandages, often reinforced with lumber and steel, have replaced rice paper and flour paste for stabilizing bones. But, overall, we use the same procedures for finding and digging bones.

Every field paleontologist has a sad tale of discovering a portion of skeleton and launching an excavation only to find that the specimen was at the end of its erosional history, not the beginning. The only bones there were the ones exposed; the skeleton did not "continue into the hill." The immense disappointment may, however, be forgotten in the quest for another skeleton, but we all want to use our time and budgets efficiently.

Our big problem, our perpetual frustration is that we cannot see into the ground. We use experience and intuition to predict more bones below or beyond the ones that lured us there in the first place. But the only way to test such predictions is to dig. No amount of wishing or dreaming can determine how much of the skeleton is still buried, locked in the rocks beneath our feet.

No rock is too hard, no mesa too tall, to deter us once we declare a skeleton important enough to excavate. Sam's bones, encased in some of the hardest rock I have ever experienced in an excavation, and buried by ten feet of sandstone cap rock, presented a real challenge. Because Sam's tail bones were articulated and the skeleton was lying in a position that indicated curvature going *into* the hillside (rather than *out* of the hill— that is, eroded away and disintegrated into fragments in the rubble at the foot of the mesa), I was convinced that excavation was in order.

Even if Sam were not new to science, the skeleton deserved complete excavation just because the bones were connected, a rare situation. Surprisingly few dinosaurs, even the famous ones, are known from reasonably complete skeletons. But here was a skeleton only an hour's drive from our museum, belonging to a dinosaur new to the state, and in an exquisite state of preservation. We launched the excavation, with outside funding assistance, and with quiet resolve to find every bone and to carry the work to a reasonable completion.

My determination was edged with apprehension as we laid out the excavation plans. Sam's skeleton might continue for fifty or sixty feet into the mesa, and we couldn't predict with any real confidence its orientation — beyond the rather safe conclusion that it indeed went in. The skeleton might be disarticulated and its bones scattered. Predicting where to dig would require me to call on all my experience. But there would also be a lot of guesswork, and perhaps uncalled-for conviction—and (I hoped) a good measure of luck.

With measuring tapes we determined the projected depth of the bones beneath the cap rock: ten feet at least and perhaps as deep as fourteen feet (depending on the orientation or trend of the skeleton). On top of the mesa, we laid out a rectangular area that I predicted should contain the skeleton. The quarry site would penetrate sixty feet into the mesa, leaving a gap thirty feet wide.

If only we could see into the ground and know exactly where to dig before commencing the excavation, we could calculate how much rock to move, the limits of the quarry, the projected costs and duration of the work. This is a paleontologist's dream.

But "seeing" into the ground had become reality for field-workers in other professions. Archaeologists had pioneered stunning field applications of technology in the past two decades in their searches for buried pyramids and pueblos. Geologists had developed sophisticated techniques to follow buried river beds in the Sahara, or to detect fault lines invisible to the unaided eye. Why not try to find buried dinosaur bones? I knew the principal difference in my plea was a matter of scale: dinosaur bones, even the largest bones of a skeleton, were two or three orders of magnitude smaller than the underground targets archaeologists and geologists had set their sights on. At best, we might expect a cross-sectional diameter of a meter for the largest bones; most would be smaller.

Sam, however, proved to be an ideal dinosaur for a set of experiments conducted by several teams of scientists from Los Alamos National Laboratory, Sandia National Laboratory, and Oak Ridge National Laboratory in Tennessee. The skeleton was buried beneath a relatively uniform sandstone, without intervening layers of other kinds of rock; it promised to be articulated and relatively easy to follow into the mesa with excavation; and the bones were among the largest known. For a smaller dinosaur, at a different site, we would surely have encountered many more variables and we would have demanded even higher resolution than needed to detect Sam's bones. These were fortuitous advantages.

In effect, I was asking these scientists as volunteers on their off-duty time from their national laboratories not only for a novel application of their expertise and equipment. I was also asking them to push the resolution of their technologies and their interpretations to ridiculously small dimensions. But the teams took up the challenge.

Chapter 4

High-Tech Paleontology

Los Alamos scientist Roland Hagan was the first to propose a working plan for using applied technology to look for more of Sam's skeleton. In his lab he had a new piece of equipment designed for just the kind of problem I posed: locating shallow buried objects. He had designed the equipment specifically for locating the 55-gallon drums that might signify hazardous waste buried in a land-fill.

Roland told me that his "ground penetrating radar" equipment had been used already with some success at archaeological sites and that it might be useful at the new dinosaur locality. He asked for details about the bone and the rock: how large were the bones? what was their density and mineral composition? what was the nature of the sandstone surrounding the bones? would the skeleton be articulated? would there be iron or other metals concentrated in the bones? I couldn't answer any of these questions satisfactorily. But Roland threw himself into the project with enthusiasm.

The device sends radio-frequency impulses into the ground; the reflections back to the surface are recorded for later analysis. The radar reflections mirror the layering in the subsurface. Under the right conditions the recorded data will indicate the presence of objects different from the surrounding matrix of rock.

I was enthralled with the idea of looking for more of Sam with this ground penetrating radar. I was also puzzled, however, because radar, as generally used, measures a time lapse between transmission and reception of the reflection. How could a sta-

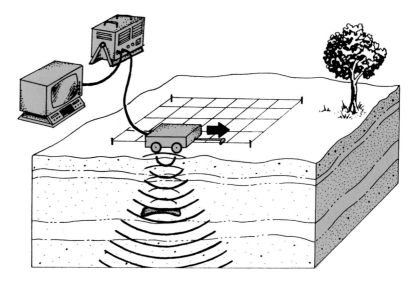

Schematic of ground penetrating radar in operation. Successive traverses along a pre-established grid produce a succession of profiles. The profiles can be interpolated to construct a three-dimensional geometry of the most likely spots to look for bone.

tionary object beneath the ground generate a time lapse? The answer became evident to me later, when we began the experiments.

Making an early site visit to see Sam's burial, or what we *hoped* would be Sam's burial, Roland also brought colleagues from the lab. Don Neeper had agreed to help Roland experiment with the radar back at Los Alamos, to learn how it operates where they could be sure of the existence of something beneath the ground. Carrie Neeper would become the record-keeper; later she would also assume the role of historian for the project. During that site visit Roland collected a sample of bone fragment and a piece of the white sandstone for analysis, to determine their densities and physical properties. The radar would be useful only if there were sufficient difference between the bone and the rock.

Preliminary results were encouraging. The bone from the Ojito site proved denser than the rock. There was now good reason to believe that the radar should work—provided the bones were large enough to detect. Roland set about organizing a trial experiment on site, with a crew of volunteers.

I had set a permanent reference point, a section of steel pipe pounded into a nearby crack in the sandstone at the edge of the mesa. This datum became the corner of our grid systems, laid out in meter squares. All trials in the remote sensing experiments would be recorded on fixed grids, and the excavation, wherever the bones would lead us, would be mapped on this reference grid. The grid mapping would allow us to compare and correlate data from all the experiments.

The day of the actual experiment, Roland and Don brought a portable generator to power the radar equipment. The sensor itself was easily portable—it was mostly an antenna and a recording device. The antenna simultaneously sent radio waves into the ground and received the reflections of those waves. The reflection data were recorded on continuous-feed paper, which produced a profile of the underground layers.

The antenna transmitted the impulses directly downward. To record the reflection profiles the antenna had to be pulled slowly across the surface of the ground. This was the answer to the puzzle: if the object being sought is fixed, then *move the radar source.* The effect is identical, because the time lapse is relative, produced from the relative difference in position between the antenna and the target. With its wheels and handle, the antenna

Antenna for ground penetrating radar (Los Alamos National Laboratory), operated by Carrie Neeper (pulling the machine) and Roland Hagan. Powered by a generator, the antenna sends radio waves into the ground.

Recorder for ground penetrating radar operated by Cliff Kinnebrew of Sandia National Laboratory. The lines mirror the layering in the subsurface.

looked more like a lawn mower than a high-technology apparatus searching for underground treasures. On each traverse along one of the grid lines the recorder produced a baffling picture of wavy lines, some pronounced, some subtle. The profiles resembled seismic records produced by seismographs.

What did the data mean? And how deep were these profiles anyway? We had no way of knowing, since we had not calibrated a known depth for reference. Ideally, each profile should have recorded reflections down to about fifteen or twenty feet beneath the mesa cap. Anything deeper would have contributed only noise. But we couldn't be sure whether these first records were on the mark, or whether they were reflecting density differences at too shallow or too great a depth.

We adjusted the scaling controls on the recorder, and worked the grid again. This round provided a more condensed profile than had the first trial and it clearly produced a record that penetrated much deeper into the ground. But how much deeper? We had records of the traverses, but we couldn't interpret them with confidence. We were learning, the hard way: by trial and error.

Roland had engineering friends at Sandia National Laboratory in Albuquerque, another research facility operated mainly for weapons development, including (then) much of the push for the Strategic Defense Initiative, or "Star Wars" technology. Wayne Cooke and Cliff Kinnebrew would bring their own ground penetrating radar equipment and their own crew-in-training to the site. Cliff and Wayne were optimistic; they routinely used radar to chart fault movements in underground vaults constructed in massive layers of salt in southern New Mexico for storage of hazardous wastes. They expected the radar to be equally effective in the search for dinosaur bones.

Rather than rolling their antenna over the uneven terrain, the Sandia engineers carried their antenna by two pairs of handles, as though it was mounted on a stretcher. Inching along a grid line in a slow and steady fashion, their antenna sent signals into the ground from a consistent height of about three feet above the surface. The Sandia team thus eliminated much of the interference produced by surface rocks and irregularities that jarred Roland's antenna on wheels.

Cliff pointed to a disturbance pattern on the recording chart as the two trainees moved along a particular grid line. The pattern curved upward through the reflection layers, then peaked and turned symmetrically downward in a parabolic curve. Having calibrated their instruments beforehand, the Sandia team, unlike the earlier team from Los Alamos, knew the depth of the layers that were represented on the chart. The parabolic disturbance pattern peaked on the chart at an eight-foot depth.

That was exactly where the upper surface of a bone two feet in diameter should be positioned. Eureka! I thought to myself, barely able to contain my excitement. We had struck paleontological gold. Rather, there was a good chance that we had. The only way to verify success was to dig. The disturbance pattern could, after all, have been generated by something else — per-

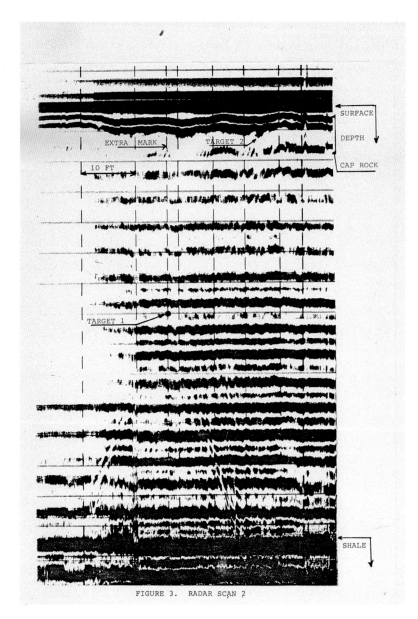

FIGURE 3. RADAR SCAN 2

haps a pocket of water-saturated sediment coincidentally at a depth of eight feet, or possibly the trace of one of the minor faults that riddled the sandstone.

Ever the optimist, I was convinced we had found bone. I wanted to cart out the pickaxes and shovels right then. That, however, was a ridiculous idea. My friends wisely persuaded me to settle for a cautious celebration of beer and New Mexico tamales.

We collected the radar profiles for more detailed evaluation. Roland, Don, and Carrie would compare the data generated by the two different experiments, Sandia versus Los Alamos, and try to predict where bones might be positioned underground.

On the same day that Roland and Don first brought out the radar equipment, they invited two other colleagues from the lab: engineer Harold Bowen and technician Rod Hardee. Harold came with a huge, floppy straw hat; Rod came with a keen imagination. Rod began almost immediately to regale us with fantastic stories about the possible meanings of the nearby petroglyphs. But they did a good job of explaining to me the kind of equipment they'd like to try on the site and the theory behind it.

The assembly of the pole and sensors for proton free-precession magnetometry operated by Phil Vergamini, Los Alamos National Laboratory. Harold Bowen, also of Los Alamos National Laboratory, built the prototype and adapted it specially for locating Sam's bones. The sensors (the knobs on the pole) receive and measure the strength of the magnetic field at that position above the ground.

My interest was piqued when Harold told me a device exactly like his had discovered the wreckage of the *Atosha*, a Spanish galleon laden with treasures that sank in the Caribbean. Harold was rather obsessed with galleons — the ultimate dream of treasure hunters, who can expect to share in the riches if they are on the discovery teams. He and Rod now became obsessed with finding Sam's bones, as if Sam were a ship buried in the ancient sands of time. The treasure they would share would be the joy of discovery; they threw themselves into the search with gusto.

Harold's device for Proton Free-Precession Magnetometry was deceptively simple. Indeed, he had built it as a prototype, without all the bells and whistles that a final version might include. The magnetometer is not a metal detector. Instead, it measures the intensity of the magnetic field of the earth at a given point. With its sensor mounted eight feet up from the ground on a hand-held pole, the magnetometer would record the magnetic field eight feet above the patch of ground directly below. Technically, the magnetometer measures the deflection of protons in response to the precise intensity of the magnetic lines of force. And it was precise, almost beyond belief: it could measure differences as small as one millionth the intensity of the earth's magnetic field.

Now, consider the earth's magnetic field emanating from points near the two geographic poles. The field lines enter the ground vertically at the North and South magnetic poles and horizontally at the equator. At middle latitudes in the northern hemisphere, the lines of force enter the earth, everything else being equal, at an inclination of about 45 degrees, slanting downward to the north. But everything else is not equal; the magnetic field varies regionally and globally over time. The distribution of continents and oceans affects the readings regionally, as do solar storms. The magnetic field can fluctuate wildly, and sometimes with great visual displays in the Aurora Borealis and Aurora Australis — "northern lights" and "southern lights." These global effects are ever-changing, dynamic, and largely unpredictable.

Variations, at an almost imperceptible scale, are also generated by local conditions. That is, the magnetic field is not perfectly uniform over the surface of a mountain, or a football field, or my back yard. The strength of the magnetic field may be concentrated, for example, above a buried Jeep in a land fill, or it may be weakened above a buried pyramid in the Egyptian desert. These variations are produced by the slight convergence of the magnetic lines of force that enter the earth when there is some buried object that attracts them. The lines of force converge toward the buried object. The more metal the object contains, the larger it is, and the denser it is in relation to surrounding rock, the greater will be the concentration of lines of force. Similarly, the closer the object to the surface, the more pronounced will be the convergence at ground level. These are the variables that produce the uneven distribution of strength of the magnetic field measured locally.

From a plot of the measurements on a grid system imposed over a given area, the strength of the magnetic field can be seen to vary. Differences may be random, or they might make a pattern with high readings clustered. These "highs" indicate only the relative concentration of the magnetic field, not the *cause* of the concentration. Ultimately, the only test to determine the cause of a set of high readings at a fossil locality or an archaeological site is excavation. This is "underground truth," a phrase that parallels "ground truth" as applied by air-borne or satellite-borne remote sensing practice. Ground truth for satellite mapping is established by direct sampling or testing, from the ground, and the results are used to calibrate remotely sensed data for extrapolation. For us, underground truth consisted of testing and sampling in the subsurface. More precisely, the underground truth lay between eight and ten feet beneath the upper surface of the mesa.

For our experiment with magnetometry, underground truth lay not directly beneath the plot of the anomaly, but instead as a projection along the angle of incidence of the magnetic field. In our case, the projection line was 45 degrees downward to the north, along the north-south line plotted as the North-South Magnetic Polar Meridian. This angle presents a problem in interpretation because the deeper the object or feature producing the anomaly, the greater the offset in the subsurface reading. For features near the surface, the angle of offset is minimal. But for features ten or twenty feet beneath the surface, the offset is substantial; predicting where to dig or where to drill a hole in the hope of intersecting a buried bone is tricky.

Harold and Rod placed small flags in the ground at every grid point, recording on the flags the readings from the proton free-precession magnetometer. They produced several scores of measurements, but all the while Harold fretted over abruptly changing atmospheric conditions: sun spot activity was then causing wild fluctuations in the earth's magnetic field, and these were detected easily by his sensitive equipment. The only solution to this variable (magnetic differences owing only to differences in the time of each measurement) would be to take all the measurements simultaneously, an obviously impractical idea unless we had a lot of magnetometers.

By afternoon Harold had recorded measurements on dozens of red flags. Although he had seen me trace the layout of the skeleton during an earlier site visit, he asked to be shown the orientation of the first eight vertebrae again, and the projection of their trend both toward the body and toward the tip of the tail if it were there.

I told him not to bother with the rearward projection of the tail because its rear half had been eroded away. I was certain. I was following the same principle I had learned in my youth as an avid fisherman: if you want to catch fish, go to the fish, don't sink your bait into barren water. Of course, the trophy lunkers tend to live in near-barren waters, since they have cleared the

Schematic of proton free-precession magnetometry in operation. The magnetic lines of force are not vertical; instead they are inclined according to latitude and local disturbances. Buried objects can cause anomalous convergence of lines of force, producing concentrations of high readings.

vicinity of all competitors and prey, but I wasn't thinking about lunkers that day.

Harold persisted. I told him he would squander his time taking measurements on the edge of the mesa. He listened politely, nodding as I spoke, but he was obviously intent on trying to envision the layout of the entire skeleton. So, I humored him; I traced a line in the sandstone where the vertebrae had been excavated and pointed beyond. Taking measurements from a region of the mesa where there were no bones might not be wasted effort after all, if it would give us calibration data.

Later that day Harold came to me with a smile and a sunburn (from reflected light) despite his big-brimmed hat. He had found bone. Where? I asked. Over there, where the tail was.

No, impossible, I replied. He showed me the flags. Two had high readings, anomalously high. You will find bone right here, he told me with conviction.

A month later we began to train new volunteers in excavation techniques. I decided to break them in by digging in the sandstone where I knew there would be no skeleton This practice in using hand tools could thus in no way injure a bone. Where better to train the new volunteers than at the tail end of the skeleton? Six inches beneath the same two red flags, the new volunteers found bone. I wished Harold had been there. When I told him by phone, neither he nor Rod were surprised. They understood Murphy's law better than I understood fishing.

Those two flags marked two very small bone fragments — each no bigger than a plum — probably from the original tail bones. We found no more bone of the tail from its middle position rearward. Incredibly, Harold and Rod had identified the positions of these two fragments with precision. If Harold were fishing, he wouldn't have needed bait; he could have snuck up on the fish and grabbed it bare handed. Underground truth.

We cannot be sure, of course, whether Harold and Rod were just lucky in placing their flags and coincidentally reaching high counts at those points. Beginner's luck, so to speak. Unfortunately, the only real test of the value of the technique for prospecting bone must be truly experimental, with controlled variables. Magnetometry holds considerable promise as a pal-

eontological tool of the future, but as of this writing, it must be considered experimental.

When Harold retired and moved to Florida to pursue his dream of finding sunken galleons, Los Alamos physicist Phil Vergamini continued the magnetometry tests in the quarry. He came with improved equipment and produced evidence of weak anomalies, but his predictions of bones never panned out because, unfortunately, his readings were made where we predicted bones but did not find any in the excavation. He was searching in barren rock.

Radar and magnetometry devices were not the only detection tools brought to the site. Another team decided to exploit the uranium concentration in the bones. Uranium occurs naturally as isotopes. U-238 and U-235 are inherently unstable. These atoms change through time by the loss and gain of atomic

The two bone fragments found by magnetometry. Harold Bowen and Rod Hardee located these fragments with the magnetometer. Both fragments come from the distal part of the tail, which had mostly eroded away. The two fragments have been exposed around their margins, where we looked (in vain) for more buried bone. The square is one meter on edge.

particles: electrons, protons, and neutrons. The changes are sequential, so that the atom that began as uranium eventually changes to another element, finally reaching stability in daughter products such as isotopes of lead and thorium. Uranium isotopes concentrate in the chemical structure of certain minerals, such as uraninite and carnotite, the principal economic sources of this valuable element.

In low concentration, uranium also occurs in ground water, sedimentary rocks, igneous rocks, and almost everywhere on the earth. The concentrations are usually so low that only high resolution laboratory instruments can detect its presence. Fossilized trees and bones often contain surprisingly high concentrations of uranium, a fact that rock hounds use in prospecting for petrified logs. Petrified logs and dinosaur bones are both big enough to offer the possibility of sufficient uranium for detection. Indeed, during the early rush for uranium during World War II, petrified logs, and occasionally large dinosaur bones, were a principal source for this precious strategic element.

According to stories I have heard but have never seen documented, the dinosaur bones of the Morrison Formation were the early source for uranium at sites and eventually in mines throughout the West — especially in Utah, Colorado, and New Mexico. Suggestively, a uranium prospect site occurs on the jeep trail leading to the Seismosaurus site. Quite likely, Geiger counter readings indicated a high at that point, perhaps from a bone. There and elsewhere the bones are often much more enriched than the surrounding rocks, making mining difficult unless the bones are highly concentrated. Fortunately for paleontology, this is a rare situation.

The occurrence of uranium in dinosaur bones in the American West seemed incidental from the academic perspective of paleontologists. No one had ever taken a direct interest in unraveling the nature of the chemical history that produces such astonishing concentrations: in some dinosaur bone, the uranium content is 10,000 or even 100,000 times richer than in surrounding rock. Quite obviously, something happens after burial of a skeleton that promotes the uptake of uranium from the ground. Bones of living animals do not contain the minerals uraninite or carnotite, so the uranium atom must be involved in some chemical reaction that binds it to the buried bone.

Even after the uranium becomes locked into the fossil bone, or onto its surface, the isotope continues to disintegrate. As a product of the decay, extremely short wavelength electromagnetic radiation (called "gamma ray emission" or "ionizing radiation") is discharged. Gamma rays are a form of electromagnetic radiation with great penetrating power; they have a frequency greater than X rays. Gamma rays generated within the ground can pass into the air, or into surrounding rock where their energy may be absorbed. Under the right conditions, gamma rays in significant abundance might indicate the presence of something underground with a high concentration of unstable isotopes.

High-energy gamma rays can ionize other atoms; that is, they can knock electrons out, creating ions. The appropriate instrument for measuring the ionizing radiation of gamma rays is a

Portable scintillation counter. The device is measuring radiation in a core that has been drilled and removed from the bedrock.

scintillation counter—which is an apt description of the bursts of energy detected by this instrument. Most scintillation counters are installed in laboratories where samples are brought for analysis. Some, however, are portable: the instrument can be taken to the sample. Naturally, Los Alamos geologists had such equipment for field investigations, and quite naturally as well, many of the Los Alamos scientists took great interest in the uranium in Sam's bones.

Laboratory trials with several scintillation counters on samples of Sam's bone confirmed that gamma rays were detectable, and that the surrounding sandstone emitted practically no radiation. With these test results, we concluded that scintillation counting should detect buried bone at the Seismosaurus site. This straightforward conclusion, drawn from straightforward tests on samples of bone and rock, seemed encouraging.

In the field the scintillation counter would work something like Hollywood's depiction of a Geiger counter: search and search, scan the surface of the ground, until the counter "sings" or the dial indicator goes to the red line. Shout Eureka! and stake a claim: we've struck it rich with uranium.

Los Alamos scientist Bill Johnson came to the site to try a "germanium high-resolution gamma ray detector"—a high technology version of a scintillation counter. This battery-powered instrument weighed only a couple of pounds. Switch it on, adjust some dials, and take scintillation readings at grid line intersections. The result: no discernible "highs" that might indicate concentrations of radioactive materials beneath the surface.

What did these unremarkable readings mean? Perhaps they meant there was no more bone on site. Or, perhaps the instrument was not functioning properly. Or, maybe the bones were too far away from the surface of the ground, attenuating the gamma ray energy so that the scintillation counter would not pick up the radiation.

Bill tested the counter with some bone fragments still in place. It worked properly, producing the expected readings. One conclusion was unavoidable: because the scintillation counting produced no significant concentrations of radioactivity, the results could not be used for comparison with either the radar data or magnetometer data.

At this point, Bill and colleagues redesigned the experiment. They assumed that the encasing rock absorbs the gamma rays from the bones, producing a halo of radiation, with decreasing intensity outward. At some distance from the bone (the source of the radiation), the intensity would be so low that the radiation would no longer be separable from the background radiation. That distance must be less than eight feet because from the top of the mesa the first set of readings did not locate any radiation highs.

However, now a better question could be formulated: through how much rock must the gamma rays pass before they would be absorbed and therefore not recognizable on a scintillation counter positioned at a greater distance? A matter of inches, or feet? Furthermore, we recognized that perhaps not all of the bone would have the same radiation levels as the ones we had measured; some bone might have much more, some much less.

To conduct this test, the scintillation counters would be suspended from the top of the mesa down vertical holes that had already been drilled for another purpose. Measurements would be taken at prescribed intervals. At depths between eight and ten feet, if the holes were close to bones, gamma radiation should be strong enough to register on the "down-hole" scintillation counter.

Variations on this promising application continued through the 1992 field season. Other dinosaur sites in the Ojito Wilderness Study Area have considerably stronger gamma radiation. For skeletons that lead into gentle slopes where the unexposed bones are shallow, the chances are high indeed that the full extent of the skeletons might be detectable by gamma ray detection prior to excavation. To date, however, this extension of the experiments has not been conducted.

Meanwhile, a fourth and final test of technological prospecting was getting underway. Through Roland Hagan, scientists at Oak Ridge National Laboratory in Tennessee learned about the

Long-period scintillation counter mounted on a tripod. By allowing the counter to record one position for fifteen minutes or even a half hour, sensitivity is enhanced.

experiments in new technologies for paleontology that were underway in New Mexico. Mechanical engineer Alan Witten talked to Roland by phone, then to me. Alan had an idea, and he needed to know the density of the sandstone, the density of the bones, their projected positions beneath the mesa. He focused on the three-dimensional geometry of the skeleton in the subsurface and the geometric configuration of the sandstone that encased the bones.

Alan's idea was to prospect for bones using "acoustic diffraction tomography." This would require vertical bore holes placed at strategic positions, drilled to a depth of at least twice the projected depth of the bones, and deeper if possible. The other techniques did not disturb the ground, but Alan's would be intrusive.

Acoustic diffraction tomography takes advantage of the differential rates of sound waves traveling through various kinds and layers of rock in the subsurface. Each rock type, according

Reducing the scatter of background radiation. Background radiation made interpretation of scintillation tests difficult. The Los Alamos team therefore tried shielding the surface being measured with a stack of protective lead bricks. Moving the bricks proved to be back-breaking work, reducing enthusiasm for this refinement.

to mineral composition and density, transmits sound waves at a characteristic velocity. Substantial differences in density — for example, the difference between low density (slow velocity) sandstone and high density (high velocity) fossil bone — can be detected by the highly sensitive microphones used in this technology. The bones are dense because of the in-filling of minerals in their cavities and pore spaces subsequent to death and burial.

Alan brought in collaborators: Chris King, a research physicist and a major in the U.S. Army, and Joe Sypniewski, then a research contractor at Oak Ridge and now a professor at Wayne State University. Together they formed the nucleus of what Alan and friends at Oak Ridge called "The Seismosaurus Tour." They even had team T-shirts specially designed.

Alan's plans required transport of the equipment by van from Tennessee. But first, core holes had to be drilled and encased with PVC pipe that was capped and sealed at its lower end. Roland offered to help find a way to get the holes drilled using Los Alamos resources.

We decided to drill the holes directly over the positions

Producing a core. The coring rig and drill bit provided by Los Alamos National Laboratory cut a hole seven inches in diameter but produced a continuous internal core with a two-inch diameter.

where radar data and magnetometry records both indicated possible "targets" in the subsurface, our euphemism for Sam's bones. Unfortunately, only a couple of grid positions had favorable readings from both radar and magnetometry tests. One difficulty was the somewhat erratic records on the grid map, a consequence of the irregular ground surface. We discovered that a planed and flat surface would have been better for record-keeping and for returning to particular points in the grid with precision. The several grid maps did not exactly coincide, which in turn created problems in comparing the data. Nevertheless, we collectively agreed to about a dozen potential positions where we might reasonably expect to encounter bone at depth.

A Los Alamos crew arrived with a truck-mounted, solid-coring drill. They maneuvered the truck over the position we rated as the best prospect. We did not expect to encounter bone at this first position to the side of the excavation perimeter. It was deliberately selected for its situation beyond the expected limits of the bone, as Alan Witten's experiment was designed to detect bone near, but not necessarily in, the bore hole. After all, if we struck bone while boring, there would be no need to use a special sensing device to know that it is there.

The drill had a hollow center. It drilled a hole seven inches in diameter, and in so doing it produced in its center a continuous two-inch diameter core for sampling. As Don Vanetten and crew brought core sections to the ground and laid them end to end on a tarp, we inspected the rock for any sign of bone. We did this first visually, searching for chips of bone or bone fragments, then with another scintillation counter, hoping at least to sense radiation from rock taken near Sam's bones. To nobody's surprise, we did not strike bone.

The drilling crew got to a depth of twenty feet. Even before we got the first hole lined with PVC pipe, in preparation for Alan's experiment with acoustic diffraction tomography, Bill Johnson took the opportunity to test for scintillation once again. He brought a different scintillation counter, a gallium arsenide detector, to the site this time. He took long-period counts (fifteen minutes at each position) at each meter of depth. For the first hole, where we expected no bone, we also

Hilde Schwartz and the author inspecting a core in hopes of identifying bone from a depth of twelve feet.

expected no high readings in Bill's measurements. And that's what he got: uniformly low radiation levels all the way down, through the expected bone level at eight feet and to the bottom of the hole at twenty feet.

The second hole was drilled where we expected to strike or be very close to bone. The coring equipment stalled at a level just beneath the projected bone level because of an impenetrable sandstone layer at that depth. When we examined the core (both visually and by scintillation) our spirits flagged. We hoped Bill Johnson might, however, find a speck of hope with the down-hole scintillation counter. But time ran out on that Sunday. Los Alamos scientists and their equipment had to return to Los Alamos and report for work on Monday morning.

Bill did not take measurements in the second hole. We even decided not to place a PVC pipe in the hole for later experiments, mainly because it was too shallow to be useful, although it did penetrate slightly beneath the projected level of the bones. Later, the hole filled in with debris.

We eventually drilled six more holes, for a total of eight — fewer than we had hoped but enough to proceed with the Oak Ridge experiments. We tested each with a scintillation detector — first on the core, then down-hole. Our spirits sank. No bone, no high scintillation counts. We each inwardly concluded that this underground truth actually *falsified* the predictions for more bone that we had drawn from the radar and magnetometer experiments. We left the site particularly discouraged. Nevertheless, we had the holes ready for their real purpose: for the acoustic diffraction tomography experiment to be conducted by the Oak Ridge team. Alan, however, wanted more holes, and deeper ones, but the six we drilled, each averaging about twenty-two feet in depth, would have to suffice.

Alan Witten, Chris King, and Joe Sypniewski packed their gear in an Oak Ridge van and drove from Tennessee to New Mexico, laden with computers, wires, manuals, video screens, sensing devices, and sundry electronic gear. By air freight they shipped the shotgun, an 8-gauge magnum mounted on wheels. It would shoot directly downward—point-blank range, we hoped, for "hitting" Sam—and its slug would deliver an impact from which reverberations could be measured at each of the six holes.

By the time the Oak Ridge team arrived we had exhausted all the remote sensing possibilities using the radar technology, magnetometry, scintillation counting, and even bone-witching tools. Few among us expected Alan to produce results any more convincing than we had seen from the other experiments. The others had begun their experiments with the boost of our own enthusiasm. Alan would have to generate his own. Alan's search would be the last detection technique tried on this site before the full excavation got under way.

The shotgun was actually a commercially produced device for sending shock waves into the ground, usually for seismic experiments. Seismologists call it a "Betsy." Alan's technique thus resembled the seismic profiling that oil exploration teams regularly use in their search for underground traps for hydrocarbons. Both use seismic shock waves, Alan producing his with a Betsy, the oil exploration teams with huge truck-mounted impact hammers called "thumpers." The acoustic shock waves, or sound waves, pass through the ground at velocities affected by the composition of the rock layers: the greater the density of the rock the higher the velocity; the lower the density, the lower the velocity.

Theoretically, because the dinosaur bones are denser than the surrounding rock, sound waves that pass through the bones should accelerate briefly, then decelerate upon reentering

the surrounding sandstone. Through equal distances of rock, acoustic waves that pass through bone should therefore arrive earlier. Variations in rock density, moisture, fault lines, roots, and untold other variables can alter the ideal transmission of acoustic waves in the ground. These variables were all present at the Ojito site, and all data had to be viewed with these nuisances in mind.

The Oak Ridge team, led by Alan, came to the site well prepared to gather acoustic data. Their Seismosaurus Tour T-shirts put us in better humor, but our pessimism was, by that time, deeply rooted. They set to work with enthusiasm, and we helped wherever we could.

To record the acoustic waves, Alan and crew reeled out a tube of wires and hydrophones. It was designed to hold a series of evenly spaced hydrophones to record sound waves at depth intervals within the twenty-foot holes. The six holes had been lined by PVC pipe and were now filled with water. The tube that

The "Betsy" in action. Peggy Bechtel pulled the trigger on the shotgun for this particular test of acoustic diffraction tomography. The 8-gauge shotgun propels a lead slug into the ground, creating a dramatic point source for generating a shock wave that is recorded by the down-hole hydrophones.

A core hole ready for acoustic diffraction tomography. The hole has been lined with PVC pipe sealed at the bottom and then filled with water. The plastic tube lowered into the hole contains a series of hydrophones, which are sensors designed to receive and record shock waves in a fluid medium. The tube itself is filled with oil. The shovel is a prop to hold the tube in position. The core hole was drilled with equipment and personnel from Los Alamos National Laboratory; the acoustic diffraction tomography equipment was provided by Oak Ridge National Laboratory under the leadership of Alan Witten. Such cooperation between individuals and institutions was common throughout the excavation.

Close-up of the oil-filled tube containing the hydrophones and electronic wiring.

contained the hydrophones was, in turn, filled with a special oil, so that the detectors would function properly in a fluid medium. Sound waves would pass from the ground through the PVC pipe, into water, then into the tube with oil, finally reaching the hydrophones which would record the time of arrival and the intensity of the vibrations.

This vertical series of hydrophones is the principal receiving device for the acoustic signals. Each hydrophone is connected by wires to electrical instruments that control the signals, and these in turn are connected to a central processing computer which receives and stores all the data for subsequent analysis. The hydrophones actually function like a hologram. With application of hologram mathematics in which the vertical orientation of the hydrophones acts as a vertical plane (hence the frequently used name for this technique, acoustic diffraction tomography), the time of arrival of the sound waves at each hydrophone can be used to calculate the position and depth of a buried object, such as bone, that rests in the plane of propagation and reception at the hydrophones.

Resolution of hidden objects underground depends on accurate measurements of reception times (to thousandths of a second and less) at the hydrophones. The "shadow" of a buried object that appears on a printout indicates its presence in a single plane and from a single acoustic source. For an acoustic source in a different position, a different shadow will appear for

the same object. Where the two shadows intersect, we can estimate the depth and position of the buried object. The dimensions of the overlapping areas of the two shadows give the crude dimensions in horizontal and vertical planes.

The Seismosaurus Tour initially took measurements from several holes, including the first and second holes. Their data was all digital; there was no visual representation of it on screen or on paper. It had to be analyzed back home in Tennessee. Even Alan hadn't a clue as to whether he was picking up anything interesting in the readings. There would be no instant gratification — or disappointment — for those of us on-site.

A few days later, I got a phone call from Oak Ridge with the results. Alan believed he had good news: he reported several targets, places in the subsurface where interference patterns indicated we could find bone. These targets were at the expected depth, and roughly in positions where we might expect bones: ten feet down and forty feet from the edge of the mesa where bones (the eight tail vertebrae) were last seen.

One of the targets was especially pronounced. Upon excavation it would prove to be the position of Sam's dorsal (rib-bearing) vertebrae. Some of the other targets would prove not to be bone, but we at least correctly established the position of the vertebral column at one point in the skeleton.

Testing Alan's predictions took two years, as the excavation proceeded inward from the edge of the mesa, following the line

Schematic of acoustic diffraction tomography in operation. The sound source, here a hammer on an aluminum plate, generates a point-source sound wave that propagates uniformly in all directions except where changes in the subsurface cause the acoustic waves to accelerate. Hydrophones in the drilled hole detect the instant of arrival of the sound waves. By computer analysis of recorded results, we can determine whether something in the subsurface (bone, we hope) caused an acceleration, leading to anomalously fast travel time for positions in which the sound wave passed through the object.

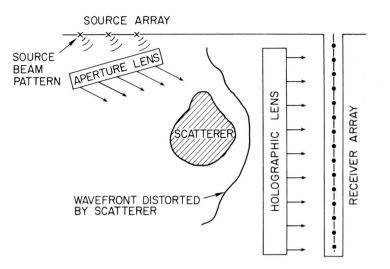

Technical interpretation of the operation of the acoustic diffraction tomography. From point sources, sound waves pass through the ground and buried objects (scatterers). If the scatterer is in the line of travel leading to the hydrophones, it may be recognized because of the early arrival times of the sound waves that passed through it.

of the skeleton. The region where he predicted bone in the subsurface proved to be the set of vertebrae between the sacrum and the neck — the vertebrae that bear ribs and thus supported the stomach and other visceral organs. Using the results garnered from the excavations, we predicted the positions of bones that we should uncover later in the excavation based on data Alan generated in later visits. Alas, these predictions proved mostly incorrect. Perhaps those false targets were water-saturated zones of rock or subsurface fault lines; they were definitely not bones.

After learning the underground truth of excavation, I came to consider the experiment in acoustic diffraction tomography

Sam's bones sensed by acoustic diffraction tomography. Excavation later demonstrated that this suggestion of hidden objects did, in fact, indicate Sam's bones. The intersection of the "shadows" mark the position and depth of something in the subsurface that caused sound waves to accelerate. The depth and approximate size of the buried object that was calculated from this data also proved to be fairly accurate.

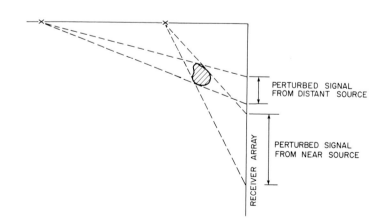

to have been moderately successful. One target, the strongest and therefore the most likely to be genuine bone, was indeed bone; the others were not, but they were not as strong. This modest success demonstrates that the technique may be useful elsewhere; applications for archaeological excavations are even more promising. Alan and the rest of the Seismosaurus Tour were encouraged enough by the experiment to publish a paper in the journal *Geophysics*. "Geophysical Diffraction Tomography at a Dinosaur Site" appeared in 1992.

Alan and his associates have lately miniaturized the equipment and improved the computer technology, making it more sensitive and more portable. And Alan has recently used his equipment at archaeological sites, where he has successfully located underground structures and tunnels with considerable accuracy. His miniaturized equipment is fully mobile and sufficiently sensitive so that the Betsy (shotgun) has been replaced with a more manageable — but less spectacular to bystanders — hammer and aluminum plate for generating shock waves. Variations on the surface-to-hole approach described here include hole-to-hole tomography, and tomography using radar waves generated by a ground-penetrating radar antenna.

The Oak Ridge experiments brought to a close the first round of testing of ground-based remote sensing technology. As I said, I regarded the seismic technique as having been moderately successful in the excavation of Sam — and with great potential for future success on other sites. But what of the other three technologies tested?

Remember the second drill hole, where we expected to find bone and did not? We had positioned that second hole precisely where the first experiments in radar and magnetometry together indicated we should find bone in the subsurface, ten feet beneath the top of the mesa. After the remote sensing experiments that used the holes were completed, we filled in that hole and practically forgot about it.

We shouldn't have. There was, in fact, bone there. But that story will come with the next chapter. For now, the point is simply that the radar and the magnetometer had proved somewhat successful, too. The experiments in scintillation counting did not locate bone, but ongoing applications of this technique still hold promise. However, it appears that scintillation counting does not pick up gamma ray emissions from sources deeper than a few inches from the surface. If future experiments show this conclusion to be true, the potential utility of this technique will be limited.

Additional trials of each technique and subsequent variations were conducted on-site over the course of the excavation. Unfortunately, Sam's skeleton was not complete, so that much of the testing on the mesa top was over barren rock, like Phil Vergamini's application of magnetometry. In my view, all of these technologies hold great promise for ushering in a new age of high-tech paleontology. They deserve serious attention and testing in the course of excavations at other sites. Applications beyond the realm of paleontology are manifold. For example, these technologies can be used to locate such diverse materials as concealed hazardous wastes, buried buildings, and archaeological ruins.

Low-tech sensing. Roland Hagan with his bone-witching instrument (a coat-hanger).

Hi-Tech Paleontology

From this modest beginning, the quest for buried dinosaur bones using ground-based remote sensing devices will surely continue. Other paleontologists have identified areas where they would like to search for buried mammoth skeletons in bogs, and several National Parks have expressed interest in the technology as a means of limiting surface disturbances at paleontological excavations. Other applications will be developed from these techniques, too, especially in archaeology and related fields. But best of all, I am pleased that we have contributed to the technology being developed to monitor land fills where hazardous materials require surveillance and sometimes intervention.

There is one final story that came out of the pre-excavation phase of our attempts at remote sensing. When Alan Witten and colleagues used acoustic diffraction tomography to look for Sam, they were, in effect, "going seismic." The name Seismosaurus took on a double meaning. My original allusion to seismicity arose from what I imagined would have been the effect of Sam taking a stroll: the ground would shake with every step of this gigantic beast. Now, the Seismosaurus Project could also be easily remembered as the first dinosaur excavation project that had used seismic and other kinds of hi-tech paleontology.

A dowser's prediction. Prior to excavation, one dowser laid out a line of rocks that allegedly marked the position of Sam's neck and head— and even the eye (shown as the isolated pebble in the middle of the head). This area of the mesa was not excavated; however, isolated neck vertebrae found more than a hundred feet away tend to discredit this low-tech prediction.

Television documentaries about dinosaurs notwithstanding, the life of a vertebrate paleontologist is not an endless series of exciting discoveries. Like all other scientific research, advances in paleontology come at a price: financial, physical, and emotional. An excavation like that of Sam is never easy, and it isn't accomplished overnight.

In dinosaur research three-fourths of our time goes to organizational details, planning, coordination of personnel, preparation of grant proposals, budget management, report writing, and communicating with landowners and land managers. We have to be good at persuading supervisors and museum directors of the worthiness of our plans to excavate and study yet another specimen. We have to train volunteers, lead site visits or tours of the laboratory, all the while responding to memos and deadlines. The other fourth of our research time goes to reading, corresponding with colleagues, attending meetings, and actually working with the fossils. The popular image of paleontologists as educated ruffians or displaced cowboys is as misdirected as the notion that our science proceeds in an orderly and predictable fashion.

Just as we carve out of our personal lives large blocks of time for the research, so do we scrape and scrounge for support. Even well endowed museums (and these are few indeed) dedicate little funding for dinosaur research. It's expensive, time-consuming, slow to generate tangible returns, and often requires extensive travel to study other collections.

For every hour of field time in an excavation, moreover,

several hours of laboratory work are required before the bones are ready for study; for my work on dinosaurs in the Morrison Formation I estimate a ratio of 10:1. That is, for every hour of labor on-site, ten hours of laboratory work follow. Alternatively, blocks of bones encased in plaster-and-burlap will accumulate in museum storage areas for decades. The excitement of discovery and excavation fades, and the harvest of unprocessed fossils becomes a museum director's nightmare and a curator's embarrassment.

A large excavation is risky from the standpoint of professional time and institutional money. To undertake a large-scale field operation requires a personal commitment to years of work, with the expectation that tangible results, such as publication and mounted specimens, may take a decade or longer to achieve. The larger the dinosaur, the greater the commitment. Likewise, the more complete the skeleton or the more extensive the site, the greater the commitment. All told, for a skeleton of a large dinosaur that may be complete or half complete, the expectations are daunting at best.

A full excavation of whatever bones might remain of Sam thus seemed an awesome prospect. We could have abandoned the site, bidding farewell to the embedded fragment of bone. In that case, all of our work would have been completed upon final preparation of the eight tail vertebrae and publication of the results. But that one bone called to me as the sirens did to Ulysses. Perhaps it would lead to more bones, and just possibly, given the right orientation of the skeleton as it lay on the edge of the mesa, much more would still be preserved beneath the ten-foot layer of sandstone and cap rock.

I had to find out. Besides, from a curatorial standpoint, if more of the skeleton were to be found, my colleagues and I could more thoroughly describe the bones and more confidently identify this dinosaur new to New Mexico.

Following my visit to the Smithsonian Institution and other museums of the eastern United States and my seminar at Los Alamos National Laboratory that had generated so many novel ideas for remote sensing, I applied to the National Geographic Society for excavation support. The single bone fragment we

left on site in 1985, a bone measuring no more than ten square inches as exposed, proved suggestive enough to bring us this vital financial assistance. The initial grant was for a one-year excavation. Later, the Society would award a two-year extension, the Martin Marietta Corporation Foundation would supply a one-year grant for more excavation, and we would win a one-year grant from the National Science Foundation for remote-sensing experiments.

In early summer of 1987, supported by the National Geographic Society, we reopened Sam's quarry with great optimism, but with a tremendous sense of foreboding, too. By all our calculations, the skeleton would lead into the mesa, below a ten-foot layer of cap rock that would be as hard as concrete. On the other hand, that one bone fragment might lead to nothing: it was possible (but not likely by my reckoning) that we had found all the bones we would ever find, and that the grant would be wasted on digging in barren rock.

Location of bone fragment. The isolated bone fragment that we left on site in 1985 was buried in the sandstone ledge in the foreground. It so perfectly matched the color and texture of the rock in which it was embedded that there was little danger of vandalism or unscrupulous collection. This fragment was evidence that we might find additional bones at the site, and it was thus vital for securing our first excavation grant from the National Geographic Society.

The first attempts by Los Alamos and Sandia scientists to locate buried bone had been provocative. We had potential targets in the subsurface, but we could verify them only by excavating. There might be nothing to excavate, or the bones might go so far and so deep into the hill that the award would be insufficient. Nevertheless, in this particular excavation, as in many dinosaur excavations, we were blessed with a crew of hearty volunteers that made it possible to stretch grant dollars much further than in other forms of scientific research.

We set to work once again with jackhammers, picks, chisels, and shovels. I predicted the skeleton would turn gradually to the north and remain at the same level. With that seat-of-the-pants projection, we laid out the area where we should excavate. Like warriors armed with toothpicks, we began our assault on the mountain of rock.

Meanwhile, the new Museum of Natural History in Albuquerque had opened in early 1986, and we were overrun with visitors and administrators. The public loved the wizardry of the state-of-the-art exhibits, and we were ill-prepared for the masses that crowded the museum halls. Volunteers were essential.

Peggy Bechtel became the museum's coordinator of volunteers, one of the most harrowing jobs in a museum. She trained and managed scores of volunteers, leading them through museum orientation, formal instruction on museum exhibit themes, details of the museum exhibits, and museum ethics. She trained volunteers to serve as docents and for behind-the-scenes labor such as carpentry and exhibit production. Her volunteers immediately became indispensable. They doubled and tripled the museum's work output. Peggy's organizational talents made the difference between opening with a mediocre museum and opening with a spectacular start. More importantly, during the next two years her volunteers contributed thousands of hours of unpaid help to the museum's operations. They became the life force of the museum.

Many of the volunteers wanted more than indoor museum work. Some hoped to excavate fossils, especially dinosaurs. After all, a "save the dinosaurs" sentiment generated the funds that built the museum. We thus had a ready and willing work force for excavating Sam.

Between the initial excavation of Sam's eight tail vertebrae in June of 1985 and the museum's opening in January of 1986, a special group of New Mexico volunteers rallied around paleontology, hungry to participate in fieldwork and research on fossils. These dedicated people organized fund-raising efforts to support an international conference on dinosaur tracks. They assisted with the excavation of an eight-ton block containing dozens of skeletons of Coelophysis, a small bipedal predatory dinosaur from the Triassic, which is now on display in the Ruth Hall Museum of Paleontology at the Ghost Ranch Conference Center in northern New Mexico. And they provided the labor for scientific study of dinosaur tracks at Clayton Lake State Park in the northeastern part of the state. Eventually they organized as a separate support group, the New Mexico Friends of Paleontology. This group is now a nonprofit corporation whose activities are dedicated to the support of paleontology at the New Mexico Museum of Natural History.

The members lent a camper shell for shelter and storage at the site of Sam's excavation. And, most important, during the first two years of excavation they contributed their time. The early stages of excavation would have been impossible without them.

Among the most avid volunteers was Wilson Bechtel, Peggy's husband, who was near retirement from his job in a local movie theater. He and Peggy together assumed full excavation responsibility in the third year of the project, under the sponsorship of the Southwest Paleontology Foundation, after I moved to Utah. Their dedication and experience during this time introduced them to the world of professional paleontologists. Their contributions as professional-level researchers in this project have culminated in several published papers as co-authors with me.

So, after a frustrating delay in finalizing the National Geographic Society grant, and after training our first group of volunteers at a nearby dinosaur site, we set to work at last. Fully two years after Frank Walker had shown me the string of bones on the edge of this remote mesa west of Albuquerque, we were finally digging in.

The bone we left in the quarry as a marker proved important for orientation, marking the level of the skeleton and its position on the side of the mesa. But it was not a vertebra. Instead it was a downward projecting bone, called a chevron, positioned at a joint in the tail. Most vertebrates with long tails, including many mammals and all dinosaurs, have chevrons. These bones protect the venous blood vessels that return blood to the body from the tail and they separate the muscles of the lower half of the tail.

But after that chevron was removed the next move was not obvious. Where we expected to find the continuation of the tail vertebrae lay only barren sandstone. The realization that the tail might not continue into the side of the mesa became a recurrent nightmare, an unstated fear that we might not find any more bones and the National Geographic grant would prove fruitless.

But within a month we did find more bone — a few yards away from the bone fragment and in the same line and orientation as the original eight vertebrae. This new bone proved to be the seventh vertebra of the tail, and we soon uncovered the edges of the sixth, and then the fifth, still tightly connected. We seemed to be approaching the pelvis.

Our hopes revived, we bore down with intensity. The vertebral column turned slightly toward the mesa, as I had predicted from my expectation that the skeleton had arched in a rigor-mortis curve before the carcass was buried. Our spirits soared, despite the hard labor. In some places, especially near the bone, the sandstone was hard and unyielding. Our quarter-size jackhammers and shovels gave way to hammers and chisels as we carved the sandstone away from the surface of the bones, following them into the edge of the mesa. One vertebra led to another, and we were immediately encouraged. Maybe the skeleton would be articulated, and (hoping against all reasonable expectation) just maybe it would be complete all the way to the neck and head.

By late summer we found ourselves facing the full ten-foot wall of sandstone that buried Sam's skeleton. I was pleased,

Discovery of more vertebrae. Subsequent excavation revealed a continuation of the vertebral column from the bone fragment left embedded in 1985. The four vertebrae beneath the square-meter scale are encased in plaster and burlap for protection during removal and transfer. These are caudals 4 through 7. In the foreground caudals 1 through 3 are partially exposed, showing the rigor mortis curvature of the vertebral column that made the skeleton turn "into the mesa."

however, knowing the National Geographic grant money would be fruitful. But looking at that formidable rock, I felt a special bond with my namesake, David, facing Goliath.

The tail vertebrae were so tightly connected that I could now entertain my wildest dream: that the skeleton might continue in this state of articulation to the pelvis and beyond, to the rib-bearing vertebrae and the neck. Except for the missing end of the tail (which would have eroded off the edge of the ledge thousands of years ago), we might have a complete skeleton. Now that's optimism, but without X-ray eyes to look into the

Close-up view of the four vertebrae (caudals 4, 5, 6, and 7). The bones are lying on their right sides.

rocks, and with the remote sensing tests only in the experimental phase, we could do little else but hope and dig.

Sam's bones presented an unusual problem. Two features of their preservation made our work exceedingly difficult. First, the bones and the surrounding sandstone were the same buff color, an unusual condition for dinosaur bone. Generally, fossil bones are dark, even in light-colored rock, and almost always they are easy to distinguish from the host rock. Not so with Sam's bones. The colors are so perfectly continuous that we had to dedicate considerable time to training volunteers just to differentiate bone from sandstone. And even professionals could be fooled. Working at the site alone one day, I spent several hours chopping gently with hammer and chisel through what I thought was sandstone when I suddenly realized I was well into the neural spine of a vertebra. We all had trouble with the lack of color distinctions.

Second, the contact, or boundary, between bone and rock

was gradational rather than abrupt as is usual for fossil bones. Before Sam's skeleton was buried, the bones had checked and cracked as they desiccated in the Jurassic sun. As sand settled around the skeleton on the sand bar, individual sand grains worked their way into the internal fabric of the bones, beneath and between the checks and cracks on their surface. One hundred and fifty million years later, distinguishing bone from rock was sometimes impossible, and often we worked in fear that we were chopping right through fossil bone. Some volunteers insisted on working well away from the bone, preferring heavy labor such as shovel-and-wheelbarrow to the close-in work with hammer and tiny chisel.

Progress became at times almost imperceptible, even after hundreds of hours of work on-site. Satisfaction came only with persistence and long-term perspective, our only antidotes for the frustrating tedium of separating bone from rock. But the reward was there bit by bit, as we followed the skeleton, uninterrupted, into the mesa.

By October, 1987, the end of the field season, we had followed the tail forward to the sacrum, which consists of five enormous vertebrae all fused into a huge boxlike structure for transferring the weight to the hind legs. Pelvis bones, too, were attached. The left ilium, oriented upward as the skeleton lay on its right side, had been eroded slightly before this region of skeleton was buried. But the right ilium on the underside of the skeleton was intact, still fused with the lateral projections from the sacral vertebrae. Three of the four lower pelvis bones (the left ischium, right ischium, and right pubis) were in place, too, but separated and collapsed from the position in life. Quite unlike the upper surfaces of the caudal vertebrae and sacrum, the ischium and pubis lay in unconsolidated sand, almost soft enough to sweep away with a brush.

Discovery of the pelvis bones was a tremendous find, and it seemed to confirm the value of one of the remote-sensing experiments. The sacrum lay beneath one of the targets indicated by the first ground penetrating radar experiments.

Removal techniques have changed little in the past hundred years. To extricate a large fossil bone from the ground requires nothing more than labor, some cheap materials, and considerable ingenuity. No one has produced a "cook book" manual on how to excavate a skeleton — probably because no two excavations are alike. Each skeleton, and each bone, presents special problems to the excavators. Keeping the bones intact is the primary concern: once freed from above and on their sides from the rock in which they have been encased for millions of years, fossil bones naturally expand. They crack and break along fractures that were microscopic during burial but which open widely when the bone is no longer confined by rock. Bones exposed on the surface of the ground for a long time almost always fracture and split into small chips and splinters; fossil bones only infrequently erode from their surrounding matrix and lie on the surface, unbroken and whole, like the string of Sam's tail vertebrae exposed by natural erosion.

Today's paleontologists use a technique invented by our nineteenth-century counterparts to stabilize bones for removal and transport. To keep them from expanding and falling apart, we apply wet tissue paper, then wet newspaper in layers, onto the surface of the exposed bone to function as a separator and cushion, then strips of burlap soaked in wet plaster laid over the contours of the bone and rock to lock everything, bone and matrix alike, tightly into place. Just as emergency-room doctors use plaster and gauze to set a broken bone, locking it in place to allow healing in the correct position, so do we apply a form of first aid to fossil bones. Perhaps "last-ditch assistance" is a more appropriate metaphor.

The product is a block of bone and rock tightly bound together in a plaster jacket. Properly applied, the jacket encases the block on all sides. If padded and secured, blocks can be transported from an excavation site to a museum or laboratory with little fear of damage. As the plaster-and-burlap bandages do not seal the contents from air, the bone and rock dry gradually, thus preventing the growth of mildew and fungus. Well-excavated bones locked in plaster jackets can be stored in museums indefinitely, awaiting laboratory preparation. Some plaster jackets are never opened, the energy and optimism of the excavation team having been captured by more important discoveries. Some museums store hundreds of unopened plaster

jackets because excavation support is much easier to secure than laboratory support.

Because the bones of Sam's vertebral column were articulated, excavation was a special challenge. Articulation is, of course, the optimal condition of preservation from the standpoint of descriptive anatomy, but it makes excavation far more problematic. Keeping the vertebrae in articulation became an important goal. We wanted to study the position of the bones in their actual orientation, connected at the joints as they were in life. Many of our crew of volunteers had participated in the excavation of the eight-ton block of Coelophysis skeletons from the Ghost Ranch quarry, so removing Sam's bones in one-ton or two-ton blocks was well within their range. One ton is still a formidable size; such a block is as heavy as a small car, half the weight of a small elephant.

When we uncovered the top surface of the first four caudals (numbers 4, 5, 6, and 7 as we later counted back from the hips),

Orientation of skeleton. Wooden models (background) mark the position of the original eight tail vertebrae excavated in 1985. On discovering the continuation of the vertebral column where these excavators are working, we established the orientation and trend of the tail, lying on its right side. The left bend in the trend of the tail (the rigor mortis arch) is barely evident at this stage of the excavation. Wilson Bechtel, left; unidentified volunteer, center; Peggy Bechtel, right.

Wilson Bechtel excavating a block for removal. The 1,500 pound block of four vertebrae had to be undercut to free it from the sandstone and then encased in plaster-soaked burlap reinforced with lumber. The operation is as dangerous as it appears in this photograph and should be undertaken only when supervised by an expert.

I decided we should take these out in a single block, keeping them together end-to-end for laboratory preparation later. The centrum (or main body) of each vertebra lay in sequential contact with the centra of adjoining vertebrae — like a series of tin cans on their sides, aligned top-to-bottom. Quite remarkably, the contacts had remained in perfect orientation: each bone had not settled separately in the sand before burial. I expected that we would be able to remove them as a single and manageable block within a week or two after commencing the downward exposure. The block would weigh only about three hundred pounds, I thought. It would be easy to handle and move to the laboratory.

I was wrong. This experience first brought home to us the immensity of Sam's skeleton. I had calculated the projected length of the entire body from the original eight caudal vertebrae, but I couldn't tell from those figures just how large the other bones would be. We dug deeper and deeper, seeking the lower limits of these four newly exposed vertebrae. We aimed to leave them perched on a pedestal of underlying rock. This is the trenching phase of excavation: finding the edges of a bone and digging straight down, to a level beneath the lowermost level of bone, in preparation for undercutting. Although I had reckoned the trench would need to be no deeper than one foot, we instead had to trench nearly three feet alongside these four bones. Their surprising dimensions more than doubled my original estimate of their diameter.

That realization was exciting. But the practical consequences were harrowing. The projected weight of the plaster jacket that could contain these four bones quadrupled: it would now weigh more than a half ton.

This change in dimensions of the block prompted us to reconsider the initial plan. Should we split the block into two manageable sections? By so doing, we could more easily handle the jackets in the quarry. We could expect to turn them over more easily in order to plaster their bottoms, and then it would also be easier to lift them out and haul them to Albuquerque. But splitting the block would surely induce more fractures through the bone. Should we dare to keep them intact?

It would have been impossible in the field to separate the vertebrae exactly along their natural joints; they were solidly fused by the sandstone, with the hardness and durability of concrete. Moreover, the bone in the joints was nearly indistinguishable from the rock. The only way to separate the vertebrae would be to drive a chisel into a crack and allow fractures to open. These fractures, however, would surely pass *through* bone rather than *between* bone. We had already used this technique of breaking through bone to remove the eight tail vertebrae from the initial discovery phase of excavation. We didn't

Exposure of remaining caudal vertebrae. The vertebrae (caudals 4 through 7) beneath the square-meter scale became Block A. The continuation of the vertebral column (caudals 1 through 3) in the foreground became Block B. The dimensions of the tail vertebrae closest to the pelvis surprised us: the side-to-side dimension of each vertebra was nearly a meter, and the height exceeded a meter.

want to break the bones just for convenience. We decided to reinforce the entire, four-bone block with lumber and steel and to find a way to lift it from the quarry intact.

Simultaneous with the trenching around these four vertebrae ("Block A") other crew members continued work on exposing three more in succession. We learned later they were the first three (and largest) of the tail. Caudal vertebrae numbers 1, 2, and 3 became "Block B." Still in articulation and lying on their right side, they curved upward in a tight rigor-mortis arch and connected directly to the vertebrae of the sacrum. The next decision involved a knotty problem: how to separate Block A from vertebra no. 3 in Block B?

We considered cutting through the rock with a chainlike device equipped with sharpened teeth, similar to a chain saw used for cutting trees. This option offered control over the separation between the two blocks, and the edges could be matched upon completion of laboratory preparation. However, sawing through the bones would leave a half-inch gap that could never be replaced. As attractive as this idea was for reasons of controlling the separation of the two blocks, I couldn't tolerate even a half-inch gap. Instead, we devised a more complicated procedure that would generate breaks in the bones preferentially along already existing natural fractures. These could then be matched and repaired in the laboratory. Block A would weigh at least half a ton.

We would begin this multi-step procedure by deepening the trenches around Block A on three sides, leaving its forward-most centrum (and the overlapping spines that interconnect above each centrum) attached to the centrum of the first vertebra (no. 3) in adjoining Block B. This would produce a pedestal beneath the bones that we could then slowly chip away. By adding props (firewood turned on edge, with shims), we could support the block underneath, without moving it, as we carefully undercut it. We knew this would be a rather hazardous operation because we had to be partly under the block as the undercutting progressed. With every several inches of undercut, we would add new layers of burlap and plaster to hold the bone and rock in place, add new props, and then resume chipping away at the pedestal.

We custom-designed an A-frame support for a hoist, to be constructed from six-inch diameter drilling pipe used for oil wells (cost for materials and welding, about $100). With a chain hoist mounted from the cross beam situated over the quarry, we would lash the block with a chain for support in case the block should fall during the undercutting stage. This would be for safety, and for stability later, if we should have to break it away from Block B.

Thus far, the plan was working beautifully. Trenching and undercutting progressed exactly as we envisioned, albeit slowly, taking the better part of two months to complete. Once the last of the pedestal was removed and the plaster jacket completely encased the block except for its contact with Block B, it was ready to move. Tightening the chain supports from the hoist, we began to remove the props one by one, expecting the block to settle slightly in the process and a break to develop naturally along the contact between Block A and Block B. Our strategy was to control the break by controlling the fall of Block A, and then to immediately plaster the broken ends of both blocks to lock broken bone into position.

Tension mounted as we removed the props. Wilson Bechtel checked the hoist and cinched it up a notch, expecting it to cushion the fall and allow us to control the direction of movement when it finally gave way. But there had been no dress rehearsal for our new and untried technique. The block was supposed to settle and break away. We were astonished when the last prop came out, and the block remained suspended, supported entirely from its side-connection with Block B. On reflection we now realize this phenomenal support of a half-ton lever system was possible because of the exceptionally strong and continuous cementation of the bones, which acted like steel reinforcement bars.

This turn of events and the profound silence it produced weren't in the plan. We replaced two props for safety and our group of fifteen excavators held a powwow. We faced the same problem as before, how to separate the two blocks, only now the risks were higher.

No saws, I insisted. We had to use breaks that could be perfectly matched and repaired in the lab. Wilson resumed

Block B hanging from the hoist managed by Wilson Bechtel. The A-frame and hoist allowed us to use a controlled fall of the block to free it from the end of what would later become Block C. The vertebral bones are not evident here, but the shape of the vertebrae in cross section is approximated by the exposed side of the block facing the camera. This face had to be covered with plaster and burlap, and reinforced with lumber, before it could be hauled away from the site.

control of the hoist, and I climbed atop the overhanging block. It didn't budge. I jumped a little, then higher. Still it didn't budge.

With a small sledge hammer, I pounded a small, cold chisel into the position where we wanted the blocks to separate. I could feel the resonant vibrations in the block with each blow of the hammer. With all my senses intent on Sam's tail, I struck the chisel again and again, driving it deeper into the rock and bone. Throughout, I was poised to jump to safety if the block should suddenly give way. Wilson strained, more from anxiety than labor, and I deliberately drove the eight-inch chisel deeper and deeper.

Suddenly, it happened. A vertical crack opened and I jumped by reflex more than thought. Wilson tugged on the chain hoist. The block fell slightly, a perfect few inches.

Wilson let the chain off a notch, so that the block's weight could fully separate the contact with Block B. It settled slightly, and then he gave a mighty heave on the chain. The block fell a little more, and twisted away from the break. We found props to steady it, to hold it still for plastering. Peggy and volunteers had paper, plaster, and burlap ready for the break, and we immediately covered both broken ends to lock the bone and rock exactly in position. Block A, suspended from the A-frame, held steady as we finished closing the jacket.

The seat-of-the-pants engineering had worked. Later we moved Block A from the quarry to the New Mexico Museum of Natural History. We began trenching Block B, which contained the three tail vertebrae closest to the pelvis. Then we began work to free the sacrum, a fusion of five enormous vertebrae, each about four feet in diameter. This became Block C. Both blocks came out of the quarry with the same procedure. Block B was slightly larger than A; Block C was the biggest. When weighed on public scales used by freight trucks, it measured 3,200 pounds. Block C was not, however, to be our largest block.

All three blocks (A, B, and C) we removed from the quarry using a truck-mounted winch provided by the Bureau of Land Management. We winched each block out on skids, then lifted them to a truck bed with the A-frame hoist. Properly jacketed and secured, these enormous blocks of bone and rock can be

Hauling the sacrum (Block C) from the site. This specially designed dinosaur truck was lent to us by the Earth Science Museum of Brigham Young University so that we could haul the sacrum from New Mexico to Utah. We were given laboratory space at BYU in which to prepare the bones so that we could make direct comparisons with the BYU collection of giant sacral bones from Dry Mesa Quarry that probably belong to Supersaurus.

transported without injury even over rough jeep trails. In days gone by, horses pulled blocks like these on wagons, or the workers crafted rails and rail-cars for transportation. Our techniques, in truth, seemed no more advanced than theirs.

As the excavation proceeded to the sacrum, the line of curvature of the vertebral column turned directly into the hill. Over the sacrum lay about eight feet of hard sandstone, and the overburden layer became higher in the direction we projected for the rib-bearing vertebrae — the dorsal vertebrae or, simply, the dorsals. By 1990, under the Bechtels' continuing supervision, the cadre of volunteers had removed enough overlying rock to follow the dorsals into the mesa. A total of seven vertebrae still in articulation were uncovered, out of a possible ten that the original carcass would have contained. Some of these dorsal vertebrae had ribs still attached in living position; other ribs had detached and collapsed to ground level before burial.

This position of the dorsals was the sole target identified by the seismic remote sensing that proved true. The Oak Ridge Team only a few months earlier had predicted that at this spot lay the best possibility for bone in the whole area. The success was rewarding.

We had proved that the skeleton did indeed trend into the hill, and with every new bone we exposed its size became more and more impressive. We proceeded to expose the dorsal vertebrae along their sides. In so doing, we began to encounter dozens of gastroliths, or stomach stones. These stones, generally the size of a plum, were purportedly swallowed by Sam to enhance the muscular grinding necessary to digest coarse plant materials. Some gastroliths were in direct contact with ribs and others were scattered away from the skeleton.

This important discovery slowed the excavation considerably, because we had to expose each gastrolith in turn, plot its position on the quarry map, label it, and take photographs. Peggy Bechtel took charge of the gastrolith excavations, and Wilson supervised and engineered the exposure of the dorsals in a huge block. At this point, about four or five workers, mostly volunteers, were on-site on a typical day. From my new post as state paleontologist of Utah I visited whenever I could on weekends and holidays, but the Bechtels had assumed the day-to-day management of the excavation.

Upon completing the trenching around the dorsals, to our great disappointment we came to the end of the articulated portion of the vertebral column. From the sacrum forward we had seven (possibly eight) vertebrae intact and still joined. These particular dorsals are sometimes called the presacral vertebrae, if one counts them from the sacrum forward. Counting from the base of the neck rearward, however, is the usual manner. Sauropods generally have ten vertebrae between the base of the neck and the sacrum. The first presacral is therefore the tenth dorsal, the second presacral is the ninth dorsal, and so on.

In mammals the vertebrae between the neck and sacrum (or sacro-iliac joint) are differentiated into those with ribs (the thoracic vertebrae) and the lower (or rear) vertebrae, which lack ribs (the lumbar vertebrae). Dinosaurs do not have this differentiation; all vertebrae between the neck and the sacrum have ribs. Thus there is no need to distinguish a lumbar and a thoracic region — hence the generic term *dorsal vertebrae*, a confusing terminology at best.

The line of Sam's vertebrae abruptly ended near the front of the rib cage, at the seventh or eighth presacral (or fourth or third dorsal) vertebrae. This disappointing termination of the series became a convenient boundary for establishing Block D: it would contain the seven presacrals and their ribs in various states of attachment. Later Peggy and Wilson discovered the next dorsal vertebra, just forward from the seven dorsals. It had fallen on its joint face and lay over a set of gastroliths.

With the discovery of that vertebra we made one more connection with the remote sensing experiments. It had to do with the drill hole no. 2 that had filled in when no bone was discovered.

On removing sandstone around the front part of Sam's torso, Peggy and Wilson came across the second core hole, filled with debris. They removed the debris, but didn't think more of it as they dug deeper around the dorsal vertebrae nearby. The right sides of these vertebrae stretched more than a meter down from the level at which they first appeared. This portion of the vertebral column, containing seven dorsal vertebrae, took on a

Murphy's hole. This was hole no. 2 in our initial selection of core-hole positions. The choice of this hole marked our best guess for the occurrence of subsurface bones, as indicated from combined data from radar and magnetometry. The hole, however, almost perfectly bisected the notch in the neural spine of a hidden vertebra, the third dorsal. That vertebra, in turn, covered a large suite of gastroliths embedded in the rock below the bottom of the hole.

Digging In

monolithic aspect, shaped like a giant loaf of bread carved out of the rock.

One morning as they started digging, with the sun at a perfect angle, Peggy peered into the cleaned-out hole no. 2. She saw bone. In disbelief she looked more closely. The hole had indeed hit bone at the lowest bone level in the quarry, about nine feet beneath the top of the mesa. That bone was a total surprise, because we thought the vertebral column had been broken apart and the forward bones had been carried away by stream action before the skeleton was buried in the sand bar.

On excavating the surprise bone, they discovered that it was the next dorsal vertebra, isolated and detached from the succeeding vertebra to the rear, and offset from its living position by a distance of four feet. The core hole itself had barely scraped the edge of the bone, and since the core sample was a cylindrical plug created and extracted by the inner surfaces of the pipelike drilling device, the scrapings of the drill did not come to the surface, and we did not see bone fragments in the spoils. More spectacular, the hole was a perfect bull's-eye: it fit almost exactly between the two projections of the neural spine, the V-shaped notch that gives these vertebrae a slingshot shape.

This vertebra, the third beyond the base of the neck and the furthest forward of all the dorsal vertebrae we would find, had fallen forward and come to rest on a pile of gastroliths. The crew carefully exposed the edges of the vertebra and the two dozen gastroliths surrounding it. Each day Wilson plotted the positions of newly exposed gastroliths, labeled them with specimen numbers, and took documentary photographs. The excavation of this isolated bone required removal of considerable rock beneath the vertebra. Much later, during laboratory preparation of the block containing the surprise vertebra, Peggy and seen in the quarry. The vertebra had protected these gastroliths from scavenger and stream action prior to burial.

This was the last dorsal vertebra found in the quarry. We found no other bones in the immediate vicinity, but later we found neck vertebrae, downstream from the torso and in the access road far from the quarry. Core hole no. 2, now renamed Murphy's Hole, and Murphy's vertebra (technically, the third dorsal vertebra or the seventh presacral vertebra) took on new significance. Our original decision to position the second core hole in a likely spot for bone, in order to calibrate the remote sensing, was a good selection after all. The underground truth came three years after the coring.

What did those sensing techniques, the ground-penetrating radar and the proton free-precession magnetometry, "see" in the subsurface when they came up with positive readings? Bone? The dense cluster of gastroliths? Both? Or something else, in which case our positioning was purely accidental? I have ruled out the last idea; our data were real and probably related to the bone and gastroliths, but whether one or the other or both, remains problematic.

We faced the same problem we had with the seven presacrals that had stayed together: how to plan, undercut, and jacket a block for removal from the quarry. The same goals applied: minimize fracturing, leave sufficient rock to hold the bones tightly in place, keep everything as intact as possible. With our success in removing Blocks A, B, and C, we were confident we could handle an even bigger block. Block D would weigh several tons, at least double the weight of the sacrum (Block C).

The A-frame and hoist were not designed to carry such heavy loads, however. We had to change tactics. This time we relied on the experience gained with the eight-ton Coelophysis

Reconstruction of the skeleton of Seismosaurus based on the known elements. Positions and anatomy of the four neck vertebrae that we collected are conjectural; these bones were isolated in the quarry and heavily eroded. Not shown in this perspective are the ribs and lower pelvic bones of the opposite (right) side, which were recovered in the excavation. The peculiar kink in the tail is based on the anatomy of the vertebrae at the beginning of the downward bend. This kink appears to be unique to Seismosaurus.

block at Ghost Ranch. We would use the same technique for Sam's dorsals.

The first part of the procedure was exactly the same. Using small jackhammers and chisels we outlined the limits of the bones, dug down several feet below the bones, and produced a trench on each side that was about eight feet long, four feet wide, and three feet deep. Then we did something new. In the trenches, the crew gently chiseled tunnels every two feet or so beneath the enormous block. We plastered underneath the block as each tunnel was expanded, placing props to support the block's underside.

Eventually we expanded the tunnels until the entire underside was free from the ground and the block was supported only by props of firewood and shims. We then fixed hefty green timbers, each about a foot in cross-section, parallel to the block along its sides. Above these we set cross timbers through the tunnels. These would become skids. Then we set new props and shims between the cross timbers and the underside of the block to lock everything into place. Now the block rested entirely on the timbers, designed like a makeshift sled. To prevent flexing, the timbers were locked together with steel bolts and braces.

Preparing Blocks A and B for removal had taken about three months each, Block C about a year. Block D required nearly two years, as the quarry was expanded and the gastroliths uncovered and mapped. Fortunately, we were able to acquire additional grants to keep the excavation going without a break and long after the initial one-year grant had been spent. Late in 1991 we hired a wrecker to pull the block containing the dorsal vertebrae from the quarry to a holding place nearby. The skids worked beautifully, and nothing in the block shifted. Finally, in 1992, Block D was moved by truck to the Museum of Natural History in Albuquerque. The museum had dedicated sufficient laboratory space for the crew to begin the long work of removing and preparing the bones.

At the public scales used by truckers to weigh freight, we weighed the hauling truck before and after delivery. This cargo must have been one of the most unusual loads ever to cross those scales. The block weighed five tons.

The articulated part of the skeleton had thus been successfully excavated by 1992, but there was plenty more to do onsite. Because gastroliths were scattered over a broad area, we had to continue to expand the quarry floor to map their occurrence and collect them for the study. This slow and tedious process required back-breaking manual labor and meticulous handwork in order to expose them without disturbing their positions.

Four isolated bones were discovered late in 1991 and 1992. All were heavily eroded cervical (neck) vertebrae that had been displaced downstream from the main part of the skeleton. Excavation of these bones was relatively routine, after the experience of the truly big blocks. Those were the last of the bones we were able to locate. All the other neck bones and all those of the head were missing, as were the smallest vertebrae of the tail. Nevertheless, we were pleased. We had found all of the vertebrae from the shoulders to the middle of the tail, all of the ribs, the complete sacrum and pelvis, and some of the chevrons.

To our great dismay and puzzlement, however, no leg bones were found with the skeleton. Trying to explain the loss of leg bones has been difficult, but the taphonomic history of the skeleton, the subject of chapter 7, offers some clues. First, let's take a closer look at the several hundred gastroliths—and what they imply about Sam's digestive system.

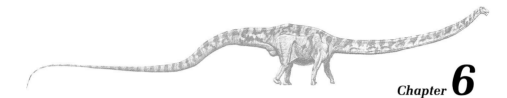

Chapter 6

Sam's Stomach Stones

Stomach stones. The phrase itself sends a knot to my stomach. I can barely imagine carrying a gut full of rocks, and I don't believe I would be much good at selecting and swallowing them.

Several times in my life I have raised chickens for both meat and eggs. These descendants of dinosaurs endlessly search the ground for tidbits of food, tidying up their surroundings. In the process, they deliberately take up sand grains for grit in their gizzard. Chicken food therefore usually includes a component of sand—essential for caged chickens that cannot forage for themselves. Evidently chickens require grit for proper digestion.

I now raise cockatiels (small parrots), an acceptable suburban substitute for chickens. Cockatiels, too, require grit in their diet. Once I forgot to fill their grit bowl, and it lay empty probably a month or longer before I discovered it was bare. I casually filled the bowl, and my four usually sedate cockatiels went into a frenzy before I could remove my hand from the cage. They were desperate for the grit, and I suppose if I had continued to deny them this ingredient in their diet the consequences would have been severe—like a conspiracy to escape.

Why do chickens and cockatiels take up grit? To aid digestion. The grit becomes lodged in the gizzard, which is a special chamber at the rear end of the stomach. With its complement of grit, the muscular contractions of the gizzard crush tough seed coatings (in the case of chickens) and perhaps woody or cellulitic stems and branches (in the case of dinosaurs) that have already been softened by chemical digestion in the stomach. Thereafter, the food passes to the intestine, the principal region of absorption and transfer of nutrition.

Seed-eating birds like chickens and cockatiels generally possess the specialized chamber of the stomach called the gizzard, or more technically, the ventriculus. This highly muscular organ is lined with a hardened horny material called koilen. The lining makes the inside of the gizzard somewhat rigid; the heavy muscles surrounding the koilen can produce considerable pressure on the contents in the cavity of the gizzard. This action crushes seeds and plant materials before they are passed to the intestine for further digestion. Gizzards are found only in seed-eating and plant-eating birds; meat-eaters and fruit-eaters do not have (and do not need) gizzards.

Most, or perhaps all, birds with gizzards use grit to facilitate digestion, although the function of the grit has never been fully explained. Apparently, birds with gizzards take up grit when it is available, but they can survive without it. Its real function is elusive. Most biologists assume that grit somehow participates in crushing the contents of the gizzard, a process called trituration by physiologists. Exactly how the trituration of food in the gizzard is improved by grit is an open question, however. Perhaps the grit rotates with muscular contractions of the gizzard, slicing and cutting into the food.

Seed-eating birds have gizzards, and birds are descendants of dinosaurs. Perhaps, then, plant-eating dinosaurs had gizzards. The similarities invite comparison, but differences between birds and sauropods suggest caution. For example, birds are toothless, but sauropods had teeth. Modern birds with gizzards eat seeds, but few (if any) Jurassic sauropods took seeds because nutritious seeds were not available until angiosperms evolved in the Cretaceous Period. Sauropods declined precipitously at the end of the Jurassic and survived only in diminished diversity and number through the Cretaceous. The rise of the flowering plants, the angiosperms, came at the very beginning of the Cretaceous. Thereafter, flowering plants expanded dramatically; they are now, by far, the dominant terrestrial plant.

Some paleontologists have suggested that Cretaceous sauropods, quite unlike many other dinosaur groups, became increasingly restricted to relict habitats left over from the Jurassic. Relicts of the Jurassic would be regions rich in ferns, cycads, and conifers. There are modern analogues: for example, the rich fir and cedar rain forests (with ferns blanketing the forest floor) that thrive along the coasts of Washington and Oregon. Perhaps sauropods would do well there. In any case, we do know that seeds of flowering plants were not available to the sauropods of the Jurassic, when these plant-eating dinosaurs reached their zenith.

There are other important differences between birds and dinosaurs, as well. Birds are small and demand rich foods for their high metabolism. Dinosaurs were large and, because their metabolic demands were lower, probably did not require food as rich as did birds.

Overall, the bird analogy is useful but rough. Are there other organisms alive today that use stones in their digestive tracts? Yes there are: crocodiles, turtles, and some lizards. According to one study, nearly all adult crocodiles of one population had stomach stones. The function of stones in these carnivorous reptiles is debatable, at least for crocodiles. Some authorities suggest that they are used as ballast for buoyancy adjustments like the lead weights that scuba divers use to make them sink. The digestive functions of stomach stones in crocodiles may therefore be entirely incidental. But at least in crocodiles, the stones are truly stones — not just small grit.

Similarly, the function of stomach stones may have been incidental in dinosaurs, too. The hypothesis that stomach stones in sauropod dinosaurs had no function (i.e., were only incidental) is as difficult to test as the obverse, but it cannot be ruled out. Sauropod skeletons sometimes contain stomach stones, but we cannot be certain that sauropods had gizzards. Nevertheless, there is considerable circumstantial evidence (especially from the excavation of Sam) that they had gizzards and used stomach stones, the dinosaur equivalent of grit. *Gastroliths* is a better term than *stomach stones* because the stones in Sam's digestive tract may not have been confined to the stomach or a chamber of the stomach like a gizzard.

Paleontologist Robert Bakker has argued for gizzards in sauropods and for the use of stomach stones as an aid to diges-

A herd of Diplodocus.

tion. His conclusions, like those of many before him, are based on largely circumstantial (and, in my view, weak) evidence, since no sauropod skeleton before Sam has ever been fully documented as having gastroliths in the visceral cavity.

Sam's gastroliths generally support Bakker's conclusions, but I am still cautious in making physiological extensions from that evidence. Nevertheless, Sam's excavation has furnished some exciting evidence that can be used to bolster the arguments of gizzard advocates. It did confirm that sauropods had (and presumably used) stomach stones, but this does not automatically mean that they had a distinct chamber like a gizzard. Absent soft-tissue evidence we cannot equate the function of dinosaur stomach stones to the presumed function of such stones or grit in birds and crocodiles. Indeed, I question the notion of any direct participation in food grinding by stomach stones in the sauropod dinosaurs, which is the function usually attributed to them. And the idea that they were used for ballast is negated by the overwhelming evidence that sauropods lived on dry land.

Sam had stomach stones — so many that they became a nuisance in the excavation. We have identified and mapped more than 240, ranging in size from about an inch to four inches in diameter, with a median size of about two inches. Most were oblong, and some were spherical. A few were flattened and roughly discoid. Each one was carefully exposed in the sandstone surrounding the skeleton. Each was photographed, its position plotted and mapped, and labeled before removal. Gastrolith documentation was thus meticulous.

Some groups of gastroliths we excavated with the surrounding rock, to keep them in position for later reference. Others were deliberately removed in contact with bones, to preserve their original positions and attitudes. Still others were unknowingly removed along with the bones in the blocks we had established to ensure the integrity of the bones. Some of these, surely, have not yet been "discovered" and won't be until the bones are fully prepared for study with the surrounding sandstone removed. With full preparation of the skeleton, we expect to find

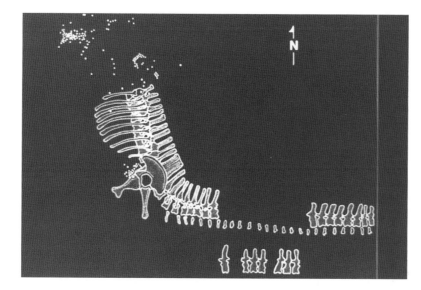

Location of gastroliths. Simplified quarry map, with ribs shown in original positions. Gastroliths were found mostly in two clusters: one near the front of the pelvis, the other farther forward. In all, more than 240 gastroliths were recovered, but not all are plotted on this map.

more, but the total will probably not be radically different from the 240 we have logged to date.

With the discovery of each stone, the progress of our excavation slowed substantially; without these stones (or with a more casual attitude as to their value) we could have completed the excavation of Sam's skeleton at least a year earlier. But this unexpected bonus in the excavation became a focus of attention when we realized that Sam's gastroliths would be the first to be fully documented for any sauropod dinosaur. And we knew that they could potentially play a role in deciphering the burial history of the carcass, that they could assist interpretation of Sam's anatomy, and that they might suggest something about Sam's feeding habits—and that of sauropod dinosaurs in general.

How can we be sure that these stones are genuine gastroliths? Skeptics, myself included, claim that most stones identified as gastroliths are river rocks that were carried by streams and deposited near the dinosaur skeletons with which they are associated. Therefore, by this line of reasoning, gastroliths may not demonstrably be associated with the dinosaurs. Their presence could be entirely coincidental, perhaps owing to the carcass acting as a barrier to sediment transport or perhaps owing to an eddying effect around the body, which would prompt an abrupt drop in stream velocity, thereby causing stones sliding along the

Close-up of the anterior region of the skeleton. The semilunar cluster of gastroliths at right center was protected from river currents by a vertebra (not shown on this map so that gastroliths would be evident) that had fallen on top of them. That bone ("Murphy's vertebra") was dorsal vertebra no. 3. Also not shown here are the four isolated and heavily eroded cervical vertebrae that were found (geologically) downstream far to the right of this map.

bottom as bed load to drop in the sands surrounding the skeleton. This abiotic explanation for the occurrence of alleged gastroliths may be appropriate in many sites where dinosaur skeletons are common. Thus, a demonstration that Sam's gastroliths are genuine stomach stones, rather than impostors deposited by streams, cannot be marshalled as evidence that all purported gastroliths are real. Instead, each occurrence must be subjected to the same test: can we legitimately rule out deposition by streams or other modes of deposition?

The gastroliths associated with Sam's skeleton surprised us. At first we casually removed them from the excavation and mapped their occurrence. Because no other sauropod excavation in my acquaintance had reported gastroliths in direct association, I was unprepared to accept the first cluster of stones that we found near the pelvis as an indication of more to come. I was naturally skeptical then, but because Sam's skeleton was buried entirely in sand, there seemed to be no other explanation.

Sam's skeleton lay in the middle of a layer of sandstone twenty feet thick, bounded below by shale and capped above by a different, more uniformly cemented and harder sandstone. Except for the gastroliths in the pelvis of Sam's skeleton, no sedimentary materials larger than sand were found in this ancient sand bar. If the stones, generally the size of a plum, were stream-deposited, there should be gravels and cobbles present in the sandstone, too, in gradational layers.

Ordinarily, stream deposits display gradations in grain size, a feature geologists call "graded bedding." It is not unusual to find gravels and cobbles associated with stream sands, but in vertical sequence. Such changes in grain size are almost never abrupt; instead, the sands are overlain by increasingly larger or smaller (clay-size) sedimentary materials that reflect a change in stream velocity and carrying capacity (bed load and suspended load). Where the change is abrupt, without gradation, either no sediments of intermediate sizes were available for transport (a rare and unusual situation), or other modes of deposition must be considered: for example, rafting of cobbles caught in floating blocks of ice or cobbles being carried to their resting place in an animal's carcass. Deposition by stream action and transport to the site in the body of a dinosaur are, however, the only two credible explanations for the stones we excavated. Other explanations, such as ice-rafting, lack evidence and need not be considered as possible.

In our quarry stream deposition seemed a very unlikely explanation for the stones. There were no gravel or cobble layers or "lenses" above or below the skeleton, nor were there any within a radius of a hundred feet from the excavation. The gastroliths in the pelvis lay directly on sand. Under ordinary depositional circumstances, stream-deposited gravels and cobbles dropped from a current would come to rest on other sediments coarser than the sand. The fact that these stones were "matrix-supported" was thus another important piece of evidence against their having been stream-deposited.

All indications thus pointed to the conclusion that the stones had been contained by the carcass, presumably in the digestive tract, that they came to rest on the sand as Sam's carcass decayed, and that the contents of the digestive tract spilled out and settled on the sand that supported the skeleton. My skepticism was dwindling. I was becoming a believer in sauropod gizzards. These were genuine gastroliths, presumably from Sam's gizzard, and we were pleased to discover that we had real documentation, for the first time, that sauropods (or at least Sam) did indeed use stomach stones.

Before the discovery of Sam's gastroliths only five sauropod

101 gastroliths. These constitute a portion of the anterior set, presumably associated with the crop. Many other gastroliths from this set were collected in situ, with rock and bone, and could not be included in this photograph. The largest gastrolith found in the entire site has been placed in the center of this photograph, partially obscured by another gastrolith. (Scale divisions are 1 centimeter.)

skeletons were reported with associated gastroliths, and the documentation in each case was meager. Three were in the Morrison Formation in North America and were only casually described. Two were in the Tendaguru beds of Tanzania, and the gastroliths were accurately measured and described. In one of the Tanzanian excavations the gastroliths were found in the neck region of a skeleton of Barosaurus. As there were only five stones, these gastroliths were little more than curiosities, but their position in the neck has an important bearing on arguments presented below that Sam had a crop as well as a gizzard.

After this first discovery of a clump of two dozen stones in the pelvic region of Sam, we did not expect to find more. The pelvis, where this cluster occurred, was exactly where a gizzard would have been in life — if the digestive tract of a sauropod resembled that of a chicken. But a year later Peggy and Wilson Bechtel, the quarry excavation supervisors, began to uncover more gastroliths — some in clusters — and by the dozens. Like the ones found within the pelvis, these stones were all matrix-supported, not associated in any way with gravels or cobbles that would have been deposited by stream action.

With each new gastrolith the map of discoveries became more and more tantalizing: the distribution of gastroliths around the skeleton did not fit the notion that they were confined to a gizzard, and we were finding many more than we expected. By the conclusion of the excavation in 1992, we had mapped more than 240 gastroliths, distributed in two general clusters: one tight group of about two dozen stones in the pelvic region, and a larger, more scattered group farther forward, ranging from roughly the middle of the rib cage toward the base of the neck. What did this mean?

As I mentioned earlier, nowhere in the quarry or in nearby sediments could we find any evidence to support an explanation that these stones were river-deposited — or ice-rafted, for that matter. We therefore held a working hypothesis that these were indeed gastroliths because they couldn't be anything else. They were brought to this site in Sam's body, where they spilled out as the carcass decayed. But this conclusion was supported by several positive lines of evidence, as well.

First, we found sets of gastroliths, including one from the pelvis and several from the more forward position, that came to rest in a line, in contact with each other as though they had been contained by soft-tissues when they came to rest on the sand. Some of the individual stones in these sets were arranged in overlapping fashion, and some were on edge rather than lying flat. These orientations are almost impossible in stream beds, except where a stream bottom is mantled by river stones. Deposition of single lines of stones in isolation seems to be nearly impossible in stream-generated sediments.

This realization that the stones actually reflected the position of stomach anatomy led to another surprise: some of these lines of gastroliths lay well beyond the outline of the carcass as indicated by the skeleton. They seem to indicate the position of entrails distributed by a scavenger that had pulled the carcass apart for easier access. The predator's tugging and pulling at the contents of the stomach brought the gastroliths along too, and they were kept intact in folds of tissue that lay on the bare sand. As the sand later filled in around the soft, probably desiccated tissues, the gastroliths were held in their original positions with respect to each other.

A second compelling piece of evidence supporting the stomach-stone hypothesis was that some gastroliths were in direct contact with bone: ribs and vertebrae. Rather than lying horizontally, some had an amazing orientation. One stone, a flattened disk that I might have chosen as a "skipper" to throw across the surface of a pond, was perfectly on edge, in a vertical position. Others were similarly oriented. Stream action could not have produced such startling positions for stream-deposited stones.

Third, measurements of these stones revealed another contraindication of stream deposits. The statistical distribution of long-axis and intermediate-axis dimensions of a natural population of river stones deposited by a surge in stream velocity would be rather linear: many small ones, fewer intermediate ones and fewer still of the very largest. The sizes of Sam's stones, however, displayed a bell curve. A few were small, the size of a peach pit. More were larger, with the highest frequency being stones roughly the size of a plum. A few were somewhat larger, the size

The largest and smallest gastroliths. The largest gastrolith is markedly bigger than all the others. Could it have caused Sam's death by choking? (Scale divisions are 1 centimeter.)

of a small apple, and a very few were larger still — the biggest being the size of a small grapefruit. Only a very unusual stream could have deposited them.

We thus were left with only one reasonable conclusion: these stones were indeed gastroliths, the indestructible dinosaur grit contained in Sam's digestive tract at the time of death. This provides a suitable standard for comparison with purported stomach stones from other sites, even where the gastroliths are not in direct association with a skeleton or their sedimentary context so easily discerned.

Satisfied that the sedimentary context is consistent with the conclusion that the exotic stones associated with Sam's skeleton are genuine gastroliths, we can deduce a great deal of information about their function, their duration in the digestive tract, the anatomy of the digestive tract, Sam's behavior in selecting suitable stones for ingestion, and the history of the carcass between coming to rest on the sand bar and its ultimate burial. (This last topic plays an important role in the chapter on forensics. Here I will concentrate on life-associated issues.)

Sam's gastroliths are all rounded, and some are highly polished. Their surface texture ranges from dull to waxy, in accordance with the degree of polishing. Except for breakage during

Gastrolith texture. Several gastroliths here demonstrate the extreme rounding and surface polishing that they undergo in the digestive system. (Scale divisions are 1 centimeter.)

More gastrolith texture. Here one of the gastroliths shows an extreme development of rounding and polish. Its composition is cryptocrystalline quartz.

excavation, none of the gastroliths have sharp edges; even the ones with irregular shapes are highly rounded. Their color ranges from white to black, with a rainbow of intermediate colors including shades of red, yellow, green, and brown. Their mineral composition is less variable: all of the gastroliths in this set originated as igneous or metamorphic rocks. All of the gastroliths, moreover, are varieties of quartz (chert and quartzite) — the hardest and most durable of the minerals commonly occurring at the earth's surface. These facts are all important in deciphering the function of the gastroliths and related aspects of Sam's behavior.

The assumption that Sam acquired these stones from local river beds is implicit here. That is the only reasonable source of exotic cobbles in the Morrison habitat. These cobbles would have eroded out of rock formations from localities more ancient than the Ojito site. Perhaps some were repeatedly subjected to the sedimentary cycle of erosion and deposition, with each rejuvenation of the landscape.

Sam most likely acquired these stones deliberately, while drinking at a stream bank. On the other hand, it is possible that sauropods migrated to source areas for particular rock compositions, an idea that I find highly unlikely but which I cannot dismiss altogether. Several paleontologists have suggested that identification of these source areas, called their "provenance," might indicate the extent of migratory movements among the dinosaurs.

According to some paleontologists, the quality of polish on a gastrolith is a distinguishing feature that separates gastroliths from stream-deposited stones. Therefore all gastroliths should have a high polish, and ones that were in residence for a long time should be waxy. The waxy trait seems to be a consequence of polish, on both the high and low surfaces of the rock. Rocks with such a texture reflect more light than rocks that have polish on the highs only, and they impart a striking quality in reflected light. They feel waxy, too, in contrast to the pitted and slightly roughened texture of rocks that lack this trait.

Not all of Sam's gastroliths, however, display a waxy surface. To the unaided eye, they range from highly polished and waxy to dull and pitted. The waxy ones are easy to distinguish from ordinary river stones, but the dull gastroliths seem not to have lost the surface texture from their original condition when Sam picked them up as river stones in an ancient stream bed.

The origin of this polish, especially the waxy texture, is difficult to explain. If these stones were used for grinding food, as generally claimed, then they should all be pitted and scratched as a consequence of grinding against each other. If the polish came from chemicals in the digestive tract — say, acids in the stomach — they should all have the same degree of polish, or at least all rocks of similar mineral composition should display similar textures.

Tumblers used by rock hounds to polish their stones are not a fair analogue for a dinosaur stomach. Stones polished in a tumbler do not polish each other; instead, the tumbler is furnished with a supply of softer polishing materials (usually culminating with talc), that impart a lustrous surface as the stones roll and tumble. Dinosaurs did not possess such carefully managed polishing materials. I surmise that these stones lay in folds and creases in the digestive tract, particularly in the gizzard (and crop), and that their polish came from the gentle action of muscular contractions.

I cannot explain why some are highly polished and others are dull. There seems to be no correlation with mineral composition, size, degree of rounding, or map position in the quarry. But I can guess, I suppose that the rocks in the specialized chambers of the digestive tract were ingested at different times, as with my cockatiels, which continually replenish their supply of grit. If so, stones in longest residence would be more polished than ones in residence for shorter times. Also, some may have been subject to polishing action more intensely than others just because of position in the digestive tract. One fold of the gizzard may have been more active in its crushing function than another fold, so that some stones in the gizzard were subjected to more of the polishing action than others.

In scientific terms we might express these observations on degree of polish as "a tendency toward high polish and eventually a waxy texture," but one must first exclude the possibility that the tendency is reversed: a waxy texture originally at the time of ingestion; rough and pitted after a long time in residence. Several experiments have documented the high polish of gastroliths; the researchers conclude that the ones with high polish, and especially the ones with waxy texture, are more polished than river stones or beach stones.

Sam's gastroliths are all rounded, a condition more striking and more universal than their polish. None are angular; none have sharp edges. In technical descriptions these stones would be described as well rounded, an extreme condition of rounding that occurs naturally in the rolling motion of rocks and cobbles in stream or shoreline settings. The mechanical action of the rolling and sliding of rocks that are originally angular (say, cubic for discussion here) wears on the corners and edges, gradually changing the shape from cubic to a cube with rounded edges and corners. Eventually the original cube approaches a spherical shape. Because of natural fractures in the original rocks from which they are derived, rocks rounded by the tumbling and sliding in rivers or on beaches tend to be more like squashed spheres than spheres and discoid, like skippers or poker chips.

Of course in nature rocks never become perfectly spherical or even flattened round by this process, but they approach that extreme: this idea is expressed as "sphericity." The sphericity of Sam's gastroliths is uniformly high. Compared with a selection of rounded stream gravels, even those preselected for round-ness, Sam's gastroliths are remarkable for their rounding.

Were Sam's gastroliths rounded by rolling and sliding in the digestive tract? Yes, but they were almost surely rounded to begin with, when Sam picked them up, probably from a nearby stream bed. Then too, when feeling the need for more stones, Sam likely exercised some degree of choice in picking them up. In Sam's digestive tract, their sphericity increased as a consequence of renewed mechanical abrasion. They were simultaneously polished and rounded. Now, all these deductions seem reasonable and they fit our expectations. But a pesky problem remains to be addressed: Why are there no angular stones?

If Sam only occasionally and randomly picked up stones, then some should have been broken and angular, reflecting the breakage and angularity of stones in stream beds. In a sample of 240 gastroliths we should expect to find a dozen or more angular stones, with sharp edges not yet rounded by the action of the digestive tract, stones taken up just before death that hadn't been subjected to mechanical breakdown long enough to reduce the sharp corners and edges. None of Sam's gastroliths have sharp, or even dull, edges or corners; all 240 are rounded. I take this collective trait to indicate that all of these gastroliths were in residence in the digestive tract for a long time, perhaps years. If so, by my argument, they stayed in the digestive tract until they were almost totally ground down, say to the size of peach pits, and passed through the digestive system with food.

Even if Sam picked up river stones in quantity, we should see a spectrum of rounding. A set collected eight months before Sam died should have more rounding (and polish) than a set picked up only a month before death; no such pattern of subsets of rounding and polish emerge from our observations. More likely, Sam picked up river stones casually and irregularly.

Considering that gastroliths are poorly documented, they have been used disproportionately in deductive arguments concerning sauropod behavior, feeding habits, and stratigraphy. All of these topics have a bearing on our studies of Sam. For example, was Sam selective? Were some river stones more attractive than others? Certainly a sauropod would select a stone within some size range (too large and it couldn't be swallowed; too small and it would pass through to no effect). Possibly a sauropod would select stones of some minimum level of sphericity. These choices could have been made by sight or by simply spitting out those that felt wrong in the mouth. But some paleontologists have hypothesized that sauropods selected for composition or color too — suggestions that, in my view, move out of the realm of science and into fantasy.

Some popular accounts of sauropods include the suggestion that these dinosaurs undertook long migrations to favorite collecting sites. There is no evidence for this proposal. I see no way to distinguish between river stones picked up in nearby drainages (but derived from sources perhaps hundreds of miles away) and stones collected from afar and carried long distances in the body cavity of a giant dinosaur. The well-intended purpose in these claims is to test the idea that sauropods engaged in long migrations; that may have been true, but gastroliths will not settle the question.

Similarly, some have suggested that sauropods selected only quartz-rich rocks for gastroliths. Again, I disagree. The fact that Sam's gastroliths (the only sauropod gastroliths thoroughly documented) are all quartz does not lead to that conclusion; instead, these may have been the only stones to survive the vigorous action of the digestive tract. Layered rocks would be easily disaggregated because of differential susceptibility of individual layers to pressure and gastric juices; rocks of other composition, such as limestones or shales, would be easily broken down in the acidic environment of the digestive tract. Such stones would not survive for long as gastroliths.

Some paleontologists have relied on moas for making projections of sauropod feeding habits and the functions of gastroliths. Moas, the huge flightless birds of New Zealand that became extinct owing to human occupation of the islands, used gastroliths extensively, and they seem to have been highly selective. These giant herbivores, weighing as much as a half ton, may have selected only white stones; perhaps only white stones were available or were the only available rocks sufficiently durable to withstand the rigors of the digestive tract. Nevertheless, I question the direct application of moa behavior to that of sauropods.

Similarly, my notions on residence time in the alimentary tract (the stones remained until virtually ground down and destroyed) have a bearing on interpretations of purported gastroliths in the Morrison and Cedar Mountain formations that are not positioned directly in association with skeletons. According to geologist Lee Stokes gastroliths occur in great abundance in the Cedar Mountain Formation (lower Cretaceous) of the western United States. This formation overlies the Morrison Formation in much of Utah and Colorado, and the two are difficult to distinguish except by subtle differences. One distinction, according to this argument, is the general lack of gastroliths in the Morrison. I agree, except that I am not convinced that the purported gastroliths are properly identified.

Whatever their origin, the widespread occurrence of these "gastroliths" in the Cedar Mountain Formation is problematic, because sauropod dinosaurs (and all other dinosaurs) are rare in that formation, whereas gastroliths are rare in the Morrison Formation, which contains the world's greatest bounty of sauropods. Paleontologist Robert Bakker has suggested that isolated gastroliths may be the only residual evidence of the positions of skeletons long since dissolved or redeposited, an idea I find attractive but hard to prove. If it can be shown that the conditions for preservation of bone were less favorable in the Cedar Mountain Formation, this idea would become more plausible and deserving of further consideration.

Sam's gastroliths have been the subject of several presentations and papers, but full description of all the gastroliths is probably years away, as the ones still encased along with bone await laboratory preparation and study.

To summarize the facts about Sam's gastroliths: they are all polished, but not all to the same extent, and some are waxy. All are well rounded; none have sharp edges. Some are markedly disk-shaped. All originated from igneous or nonfoliated (un-layered) metamorphic rocks; none are sedimentary or layered; and they range from white to black, including a rainbow of colors. Three-fourths of the gastroliths are between one and three inches at their longest dimension. Long/intermediate/short axis dimensions of the smallest are .83 inches x .67 inches x .51 inches; and the largest is 3.74 inches x 3.58 inches x 2.87 inches. Large (greater than 3.10 inches) and small (less than 1.20 inches) gastroliths are uncommon.

Not included in the summary above are observations concerning the positions of the gastroliths in the quarry. These facts are just as important as the physical qualities of the objects themselves. Contrary to expectations, Sam's gastroliths were not found in a single pile, as though dumped from a bushel basket and immediately buried. Instead, Sam's gastroliths were spread unevenly over more than 1,600 square feet. We may have missed some in the excavation, especially in the early years before we fully appreciated their implications, but the patterns of their distribution seem clear. In Sam's carcass were two clusters of gastroliths, one large set in the forward region of the digestive tract and a smaller set in the rear region found within the pelvic bones. Only the rear set, consisting of 26 tightly clustered stones, can be attributed to a gizzard, the grinding chamber originating as a specialized rear pocket of the stomach.

The set from the front of the body seems to be centered at the front part of the chest cavity, near the base of the neck. Evidently the neck and head were displaced after the gastroliths spilled out of the body cavity, leaving the gastroliths to mark the temporary position of the front part of the body. Between the front set and the tight cluster in the pelvis was a barren region of at least four feet along the vertebral column where no gastroliths were found. This separation is some evidence of a specialized forward chamber in Sam's digestive system, which I have called the crop in reference to similar anatomy in grain-feeding birds such as chickens and cockatiels.

There are no discernible differences in roundness, polish, size, or composition between the front set (from the crop) and the rear set (from the gizzard). This fact is inconsistent with the usual function of the crop as a storage chamber, where grains and plant matter collect in lumps before passing to the stomach and gizzard for digestion. The inconsistency lies not in the anatomical position, but in the fact that the gastroliths are rounded and polished, and many are waxy. If the crop were simply a storage chamber, then gastroliths should have no function there. The fact that they are highly polished and rounded, like those from the gizzard, indicates that the front gastroliths were equally involved in processing food, a radical suggestion for the function of a crop. One highly unusual bird, the primitive hoatzin of South America, has a crop with a gizzard's function. Although the anatomy of one bird is not definitive evidence that Sam had a crop, it does demonstrate that a crop with a grinding function is possible.

Overall, we have sufficient evidence to propose that Sam's digestive system resembled that of grain-feeding birds: esophagus, crop, stomach, gizzard, and intestine. The crop and gizzard both contained gastroliths, where food was probably pulverized in preparation for chemical digestion (gastric secretion in the stomach and absorption in the intestine). Yet, a nagging question remains: how exactly did Sam (and other sauropods) use the gastroliths in digestion?

According to conventional explanations sauropod gastroliths were a substitute for strong teeth. They were used in a grinding function in place of chewing. I disagree, for several reasons.

First, not all sauropod skeletons seem to have had gastroliths, but all sauropods had decidedly weak dentition. I doubt that stones as a substitute for teeth were universally necessary for mastication. In modern animals dentition is suited to diet. Only birds, which lack teeth, have been documented to re-

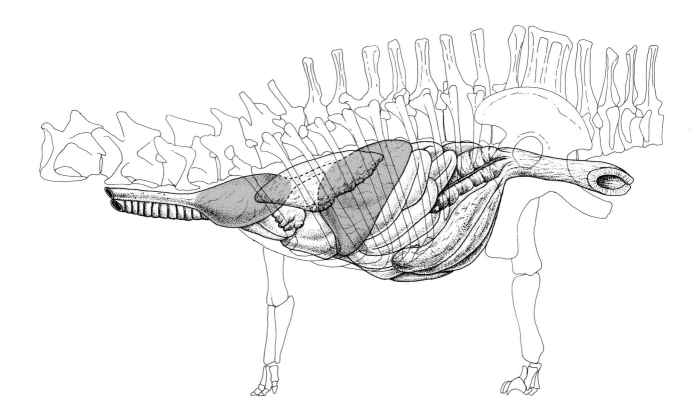

quire grit as an accessory material for grinding food—and only seed-eaters among the birds.

Second, gastroliths must have remained in the lowest creases and folds in the capacious digestive organs, whereas food would have occupied the chambers from bottom to top; only foodstuffs at the bottom of the crop or gizzard would have been subjected to a grinding action of the stones.

Third, the bulk of the gastroliths is surprisingly out of proportion to expectations: the 240 from Sam's carcass scarcely fill a ten-quart bucket. To be important as an accessory grinding device, the bulk of gastroliths should be considerably greater, or the foodstuffs must have been forced through narrow constrictions blocked by gastroliths through which the food had to pass before moving to the next chamber of the digestive tract. For an animal weighing five or six times as much as an elephant and processing many times as much food every day, the gastroliths seem a weak addition to the digestive tract.

Instead, I propose that the gastroliths stirred digestive juices as they rolled and tumbled in the bottoms of their respective

The sauropod digestive tract. This schematic includes the crop (the forward chamber at the base of the neck) and gizzard (between the stomach and small intestine).

chambers, much like a magnetic stirrer in a chemistry laboratory. A small stirring device can create sufficient turbulence to circulate fluids throughout a beaker, preventing unwanted settling and segregation of materials by density. By this argument, gastroliths had a similar role: to keep foodstuffs and the digestive juices mixed, not to grind the food or pulverize it. I subscribe to this notion especially for the crop, which was probably distensible and not highly muscular. Fermentation and putrefaction in the crop might be controlled by adequate mixing of digestive juices, preparing food for more vigorous chemical treatment in the stomach. A grinding function for the gastroliths in the gizzard seems more reasonable, where muscular contractions alone would keep the food moving and prevent stagnation.

In both places, the rounding and polish of gastroliths would arise from their contact with each other, the lining of the digestive tract, and with gastric juices, not with the food materials; thus the grinding (rounding and polishing) of gastroliths is a consequence of their tumbling and rolling with each other, with only an inconsequential grinding of food.

Interpretation of Sam's gastroliths has been a fascinating part of the excavation of Sam. The gastroliths may tell us more about how Sam lived, and whom among the living the great sauropods most resemble from a food processing standpoint. But might the gastroliths also tell us how Sam died?

I offhandedly remarked to a reporter that Sam's largest gastrolith, the size of a grapefruit, was considerably larger than all the others — and maybe it caused Sam's demise by lodging in the esophagus or trachea. This tongue-in-cheek (or, stone-in-mouth) remark received considerable publicity. Did Sam choke on this giant gastrolith? Maybe. The evidence is far from conclusive, but the possibility cannot be ruled out. There is, however, evidence suggesting another cause of death: Sam may have been killed by a predator.

Chapter 7

Forensics with Sam

Near Sam's carcass we found the tooth of a predatory dinosaur. Was Sam killed by a predator?

In our judicial system, when there is suspicion of a wrongful death, specialists, usually including medical doctors, are called upon to reconstruct the crime: How did the victim die? What were the circumstances leading up to the time of death? What happened to the body after death? These are the questions asked by practitioners of forensic medicine. We shall do the same with Sam. As paleontological coroners, our mandate to determine the cause of death must be limited to the facts, including facts deduced from circumstantial evidence.

For paleontologists the court is our technical journals, where we submit articles (technically known as "papers") to present our case to other paleontologists. Truly, we are judged by a jury of peers. We become paleontological coroners because we want to know more than just the identity of the fossil organism. We want to know how it lived and how it died.

Once we piece together as much evidence as we can, we marshal our arguments in a logical, (usually) dispassionate fashion. Papers published later by other authors may cite us pro or con. This is a kind of forensic paleontology, and the principles are no different from those applied by medical specialists involved in criminology.

The technical name for forensic paleontology is taphonomy, and the specialists are taphonomists. The goal is to reconstruct the scene where the animal died as a historical account. Development of a plausible taphonomic history has been a

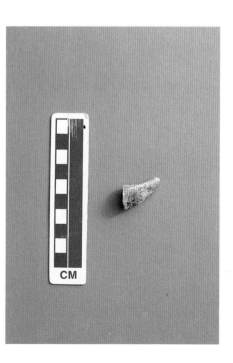

Tooth of a predatory dinosaur, found near Sam's carcass. This tooth is probably from Allosaurus or a close relative. It is the only indication of any other dinosaur at the site.

persistent goal in paleontology almost since fossils were first recognized as evidence of past life.

In paleontology, as in forensic medicine, we are often faced with hundreds of facts or observations. We're never certain that we have collected all the facts or that we have chosen the right facts to record. Sometimes we overlook something that could be a critical link in unraveling a story. Such is often the case in mystery novels. Sherlock Holmes would have been an excellent paleontologist, an excellent taphonomist; in fact, his creator, Sir Arthur Conan Doyle, was an avid amateur paleontologist. Conan Doyle expressed considerable interest in dinosaurs and dinosaur tracks in *The Lost World*. He was even implicated in the Piltdown hoax (the excavation of a fraudulent skull and jaws of a hominid) — but that allegation was made without proof of complicity.

Exposure of the Piltdown bones as a hoax impressed on paleontologists the importance of acquiring new specimens only by excavating the fossils themselves — not by purchasing them from dealers or receiving them from a personal collection. Today's paleontologists, moreover, are equally concerned with the context (sediments and associated geological features), content (the fossil's anatomy and identity), and associations (other organisms). We are no longer satisfied with the simple recovery of a skeleton. The circumstances of its occurrence are just as important as the bones, and these are lost without meticulous attention to details that transcends interest solely in a skeleton. We must be detectives at the time of excavation, because the very act of excavation destroys evidence that can never be recovered.

In my supervision of the excavation of Sam I tried to impress upon the field crew the importance of being deliberate and thorough. I told the crew, "It's not a race." We collect as many facts as we can as we excavate, because by its very nature the removal of rock and sediment around Sam's bones destroys the clues that may be critical to understanding the life and death of this particular sauropod. For Sam we have lots of facts. Reassembling the evidence is something like solving a mystery, and in this case it could involve a killing. After all, it's possible that Sam was overcome by a predator, and we even have one tidbit of evidence to support that idea—the tooth of a predatory dinosaur.

First, let's assemble the facts and their immediate implications. All observations are appropriate, including what's *not* there. For example, there is no evidence whatsoever that Sam was an old individual or that death was due to old age: no injuries, no degeneration of bones from arthritis (which is known in some dinosaurs, but not Sam), no evidence that Sam was malnourished or diseased. None of Sam's bones were broken. There is no indication of steep topography, such as a cliff or a steep wall of rock, that might have caused accidental injury and subsequent death. The potential list of what's not there could be expanded for pages, but this partial list will suffice as a reminder that one must be aware of negative evidence in weighing the possibilities.

We know Sam's identity from our studies on taxonomy (see chapter 2): Sam was a large sauropod dinosaur related to Diplodocus and Apatosaurus, and Sam lived in the same area (what is now the Colorado Plateau) and same time (late Jurassic) as these two genera of herbivores. We do not, however, know Sam's gender. There are few clues among sauropod dinosaurs that call attention to sexual differences in the skeletal anatomy. Perhaps in the future we may be able to determine the sex of an individual from chemical evidence, but that's only a dream at present. If that technological advance should happen, however, the bones will be available for study—perhaps by a researcher in another museum or in a university. Curators regularly allow sampling for laboratory analysis much in the fashion of coroners exhuming a human body from a grave to make new tests for identity or cause of death.

Sam's vertebrae were lying on their right sides, and most of them were still articulated. Some of the ribs were still attached to the dorsal vertebrae, especially the ones on the right side, the lower ribs as the body lay in the ground. Other ribs, especially the ones on the left, or upper side, collapsed before the skeleton was buried. At least one of Sam's ribs was detached and carried a couple of yards away, perhaps by a strong current, but maybe it was flipped over by a predator (if there was one) or a scavenger.

Between caudal vertebrae numbers 1–7 (which were in articulation and continuous with the sacrum) and numbers 20–27 (the original eight excavated in 1985) was a gap in the line of the tail. A few fragmentary vertebrae, too eroded to provide much useful anatomical information, were found out of line but roughly in the position of the gap. We're not sure that we found all of the ones that belong in the resulting gap. At least five from that gap were there on-site when we began excavation in 1985. They were encased in heavily eroded sandstone mounds about six feet to the south of the main line of the east-west trending tail. Others seem to have eroded away entirely, so we will never have a complete set of Sam's tail vertebrae.

Accessory bones (called chevrons) in the tails of reptiles and kin, especially large dinosaurs, occur on the underside of each joint. In Sam's tail these chevron bones projected downward and rearward, protecting blood vessels and nerves underneath the tail and separating muscles on either side of the tail. The chevrons are considerably smaller than the vertebrae, so they could have been more easily moved by current action or by a scavenger.

Some of Sam's chevrons — at least three — were still in place when we excavated, that is, they were still attached at their respective joints in the tail, small bones next to large ones. This indicates that the bones were not separated by currents to any great extent, for if they had been carried by a stream, larger bones would have settled out first, smaller ones later, and the result would be a jumble of disconnected or even widely displaced bones. Actually, the hydrodynamic properties of bone are more complicated than simple distinction by size. Their shapes are also important: compact bones such as vertebrae are less easily moved by currents than long or flat bones such as the scapula or ribs. At least one other chevron, a large one from near the base of the tail, was stripped away from its original position and was found about a yard away. Like the displaced rib, this bone seemed to have been carried downstream slightly, as indicated by the slope of the bedding planes in the sandstone.

I once came upon a sea turtle skeleton on Padre Island near the southern tip of Texas, in the Gulf of Mexico. The bleached bones of the shell were stranded in the beach sands, still partially articulated, the dead turtle having been carried there by the high surf in a storm. The leg bones were nearby, just down current from the giant shell of the turtle. They had been carried away secondarily, after the turtle carcass disintegrated. I searched in vain for bones of the skull and feet. Being smaller than the shell and legs, these bones must have been light enough to be carried away by currents or wave action. Not realizing it at the time, I was a late-arriving witness to the taphonomy of this turtle's remains. They had been partially separated by wave action, after death and decomposition. If I had searched further, I might have found the skull and foot bones, but I had not yet learned to think critically about taphonomy. Moreover, I had come upon the skeleton as a beachcomber; it was not my professional quarry.

For Sam there are no leg bones or foot bones. As the excavation proceeded, we vainly searched to uncover some of the limb elements. It became clear, however, that the bones of the feet and legs were no longer associated with the torso or tail. Perhaps they remain hidden downstream in the Jurassic sand bar, beyond the limits of our quarry—just as my sea turtle leg bones were probably downstream from its shell. But if they are too far away, and even if found, they could not then be certified as parts of Sam.

The orientation and positions of the bones we did find can tell a great deal about Sam's death. As the bones were taken out of the quarry we plotted their positions on a grid map. Sometimes plotting the bones is difficult because we don't know exactly which bones are in a plaster jacket, and won't know until months or years later, so we plot the positions of the jacket on the map for future reference. Slowly the orientation of the skeleton takes shape on the map as the excavation develops.

Several important pieces of evidence about Sam's death can be seen on the map. First, the line of the vertebrae is curved in a tight arch. The curve is unnatural, as though the tail were lifted vertically. This is a rigor mortis posture, produced as the carcass lay on the surface of the ground without being pried apart by scavengers or buried too quickly beneath confining sands and muds.

Just after death, as an animal's body cools, the muscles and tendons stiffen, becoming almost rock hard. This stiffening is what causes the arching of the back. Muscles that hold the body up, acting against gravity in life, are larger than ones that pull the body downward. For this reason the muscles on top of the tail are more massive and therefore stronger than the muscles underneath. These muscles work in opposition to each other in life. The smaller ones underneath are assisted in their work by gravity, but after death with the carcass lying on its side, gravity doesn't help. The larger muscles on top "win" out in their stiffening after death. The rigidity is caused by the contraction of muscle fibers, the larger the muscles the greater their capacity for contraction.

The arching of Sam's back after death came about in this same way. The muscles above the tail and above the ribs and hips were stronger than lower muscles. The carcass remained unconfined by sediments, lying on the surface of the ground undisturbed. As the muscles stiffened the body arched backward.

One summer while excavating Sam's skeleton, we spotted a dead cow near a fence line on the range not far from the road. It had died only recently, for we had not seen the body the day before. Every day, as we passed "dead cow hill" I scrutinized the carcass. The hot summer sun caused the skin to desiccate and turn rigid. The legs became stiff and outstretched, and the body—originally collapsed—had gradually, over a period of several days, rotated to its side. The legs projected sideways, stiffened, and lifted from the ground.

By the fourth or fifth day, the body was not only stiff but quite bloated. The most peculiar change came with the head and neck. They had gone into the rigor mortis curve. Lying on its side, the head arched steeply backward, in a contorted posture. The heavy lifting muscles of the neck, situated on top of the neck bones, were more powerful than the lower muscles. The lifting muscles dominated in the rigor, creating this altogether unpleasant orientation of the carcass. Eventually the body fell apart. Vultures, crows, ravens, and other scavengers consumed the flesh, and the bones collapsed on the hard dry ground. They were never buried, however, so they never had a chance to become fossils. Instead they crumbled in the sun, and scavenging animals reduced them to powder and food.

Another important line of forensic evidence from the map

of Sam's bones are the facts that the legs are missing, the neck is disconnected and most of it is missing, and the skull and jaws are missing. We were unable to determine where the legs are, but we do know that they are not connected to the body as they were in life. We found four neck bones downstream and disconnected from the main skeleton. They were heavily eroded from modern traffic (unknown to us, they were in the access road leading to the skeleton) and their positions cannot be determined with confidence. The neck bones we found were probably from the middle of the neck; the bones from the base of the neck, and from the head end were missing. Similarly, the bones of the head (the skull, jaws, and teeth) were missing, too.

Clearly, some of the skeleton had been separated before burial. Perhaps the missing bones are all downstream from the main body, buried in the sandstone only a few yards away like the rib, chevron, and four displaced neck bones that we did manage to find. The total area quarried is only about two thousand square feet. Every inch of that required hours of labor. We will always be haunted by the fear that just one more blow with the hammer-and-chisel on the sandstone might have exposed a missing bone.

Most of Sam's ribs are still connected at their joints on the vertebrae. The carcass was not entirely tugged apart by scavengers or torn apart by currents. Only one rib was disconnected, and it seems to have been flipped over once, coming to rest on top of the hip bones. The ribs of the upper (left) side all collapsed somewhat before burial, but the ribs on the lower side remained attached in their original positions. The rib cage had collapsed slightly before burial, but nearly maintained its original three-dimensional shape.

We have other important facts in this mystery. First, as the bones lay in the quarry, their upper surfaces (the parts facing toward the sky) are checked and cracked — not from our tools when we excavated, but from exposure after death. This deterioration is characteristic of bone that is exposed to air, like the cow's skeleton I watched deteriorate on the open range. Obviously some of Sam's bones were baked in the sun, cracking and flaking as they dried.

The bones were not completely desiccated, however. On the

undersurface of one pelvis bone (the pubis), which stayed moist in contact with the ground, are several dozen tiny bore holes made by unidentified organisms, probably worms. They may have been feeding on organic material in the bone, such as proteins and fats. At least on that surface, hidden from the sun, there was enough moisture in the ground to support decomposer organisms.

The sediments, the sands around the bone, were all laid down by water. They were subjected to rather strong currents generated by a sizable stream or river. They resemble the sands deposited on the inner curves of streams as sand bars.

The texture and internal patterns of sandstones can provide geologists with a wealth of information concerning stream direction and velocity at the time of deposition, and whether the current was persistent or short-lived, giving way to gentler currents or even quiet water. In slowly moving water, suspended particles smaller than sand settle out and fall to the bottom as a layer of silt, because the turbulence is not strong enough to keep them moving. Such deposits become siltstones. In quiet water, even microscopic clay particles can settle out of suspension, leaving a muddy surface at the bottom. These may become mudstones or shales. Around Sam's bones there was no evidence of silt or clay deposition — only sand. Stream currents in the river where Sam's carcass was buried were thus strong and persistent, with little fluctuation.

That is the evidence derived from what is left of Sam. But was anyone else at the scene? This brings us to the stray tooth of a carnivorous dinosaur, perhaps Allosaurus or a close relative. It was found three feet downstream from the sacrum. The tooth is fairly large, from a predator that weighed at least a couple of tons and resembled the formidable Tyrannosaurus.

This tooth shows that a predator was there and that Sam's carcass attracted at least one meat-eating dinosaur. The tooth could not have drifted in on a current; otherwise, similarly sized rocks and pebbles would be scattered in the sandstone too.

Dinosaurs, like modern reptiles, grew continually through their life span. Their jaws grew continually, too, and with them grew teeth. They did not have permanent teeth, like those of adult mammals; instead, their teeth never fused to the jaw-bones. New teeth grew beneath the functional exposed teeth. As an old tooth wore out, or loosened in a feeding session, the attachment in the socket weakened and the tooth would fall out. New teeth growing in the jawbones soon replaced the missing teeth. Teeth of predatory dinosaurs are often found near carcasses of their prey, a permanent record of one dino-saur's feast.

I learned about this aspect of reptile anatomy in a rather memorable way as a senior biology major at Michigan State University. On a herpetology field trip to a marshy area in the northern part of the state, Professor Hensley demonstrated with a blue racer how to subdue an excited, thrashing snake caught by its tail. An hour later I glimpsed a rapidly disappearing tail in the wet bog. I lunged, knowing it was a harmless but excitable water snake. As though prompted by our professor, it began thrashing uncontrollably, but I had a strong grip near the end of the tail. I slung it up in a broad arc, and brought it down rapidly, the centrifugal force causing this irritated female (laden with eggs) to stiffen. As instructed, my arc carried her head down-ward and between my legs. I closed my legs together, pressing her long slender body between my thighs as I pulled her for-ward in one quick motion in order to grab her head.

My execution was perfect up to this point, but I was too slow. As I pulled her head between my legs, she reached sideways and took a powerful bite on the inner side of my thigh. I was startled, for I had naively expected her to cooperate, but I was not defeated. She finally gave up the struggle as I grasped her bulging midsection and neck with a firm grip. We identified her and released her back to the marsh. Later, I found several teeth lodged in my thigh, and several others in my levis. I had learned first hand about the manner in which reptiles shed their teeth.

Now, back to Sam: from the evidence at hand, can we assert that the meat-eater *killed* our giant herbivore? No, not with certainty, because it is likely that meat-eating dinosaurs did not always kill what they ate: if they found a carcass with flesh, they surely took advantage of a fast and easy meal. So, on the basis of the tooth, we arrive at two alternatives on the cause of death: Sam was either killed by a predator or Sam died of other causes and was then scavenged by a carnivorous dinosaur. I think the second possibility is most likely. I have a hard time imagining how a two-ton predator could kill a prey animal ten or twenty times heavier. Young or weak members of the species, yes; but the magnificent Sam, no. The predator may, of course, have been acting in a group, like a pack of wolves. This idea has been presented recently by dinosaur artists, but with little real evidence that predatory dinosaurs cooperated in the hunt. Even so, the scale of Sam to an Allosaurus would have been similar to that of a lion and an elephant; even prides of lions do not attack a healthy, adult elephant.

At the very least, the tooth indicates that a meat-eating dinosaur was there and lost a tooth on site, just as the racer's teeth in my thigh indicated a snake had been there. The presence of the snake's tooth did not mean the snake had tried to eat me — nor, for that matter, I it. Likewise, the carnivorous dinosaur that owned that tooth until the time it fell out of its socket may not have been interested in feeding at all; the association of its tooth with Sam's carcass could be coincidental. What evidence might we accept as *proof* that a meat-eating dinosaur fed on Sam's flesh? Gouges in Sam's bones, or a tooth embedded in a bone might be clinchers, but such evidence is rare and we have none this convincing for Sam.

But, assuming for the moment that the presence of the tooth is evidence of a real association of a carnivore (whether acting as predator or scavenger) with Sam, one very useful induction can be made: the portion of the skeleton we discovered had not moved since the time it had flesh. That is, the skeleton was not secondarily carried away by currents after the flesh had deteriorated. For this reason, we suspect that the missing parts of Sam's skeleton could be found nearby. Even after we declare the excavation finished, which means that we give up the search for missing bones, we or other paleontologists will continue to watch the site for new bones to appear as a result of erosional processes, the same processes that initially exposed the original eight tail bones. In this way, perhaps decades from now, more of Sam's bones may be recovered.

The distribution of the gastroliths, described in the previous chapter, documents the same conclusion: the carcass came to rest here, its flesh was stripped away, and the stomach stones spilled out of the digestive tract. The bones of the rib cage and the back were not moved again; otherwise, the gastroliths and bones would not be found together. The scatter of the gastroliths indicate that a scavenger tore apart the visceral cavity, then tugged and pulled at its contents, eviscerating the carcass. With the evisceration came the gastroliths, redistributed by the actions of the scavenger as the guts were consumed or rotted. This pattern in turn leads to another speculation: the scatter seems extraordinary for a single scavenger acting alone. More likely, several scavengers competed for position on the carcass, and one or several may have tugged the entrails and pulled them apart at a safe distance from the main carcass. The skull and neck may have met a similar fate.

The gastroliths are important in our investigation for another reason: Sam may have choked to death, as mentioned in the previous chapter. The giant gastrolith crudely approximates a sphere in shape, like a contorted grapefruit. Even with alignment of its long axis parallel to the throat in swallowing, the bore required for the safe passage of this stone through the esophagus to the crop would have been prodigious in comparison with the bore necessary for safe passage of the next largest and all the others. The giant gastrolith is anomalous. Albeit an unorthodox idea, this scenario merits serious consideration.

Now, let's piece together these facts and their possible implications to reconstruct Sam's last days as a live animal, and the first days of the carcass in its long journey to becoming a fossil.

With all this evidence, we still cannot identify a cause of death. As discussed above, Sam may have been killed by a predator, or death may have come by choking on the largest stomach stone. Both ideas are plausible and can't be eliminated, but neither is satisfactory. I doubt that I could convince a jury that one or the other is the actual cause of death. Sam may well have died for some other reason.

Our forensic uncertainty is not unusual: in only a few cases

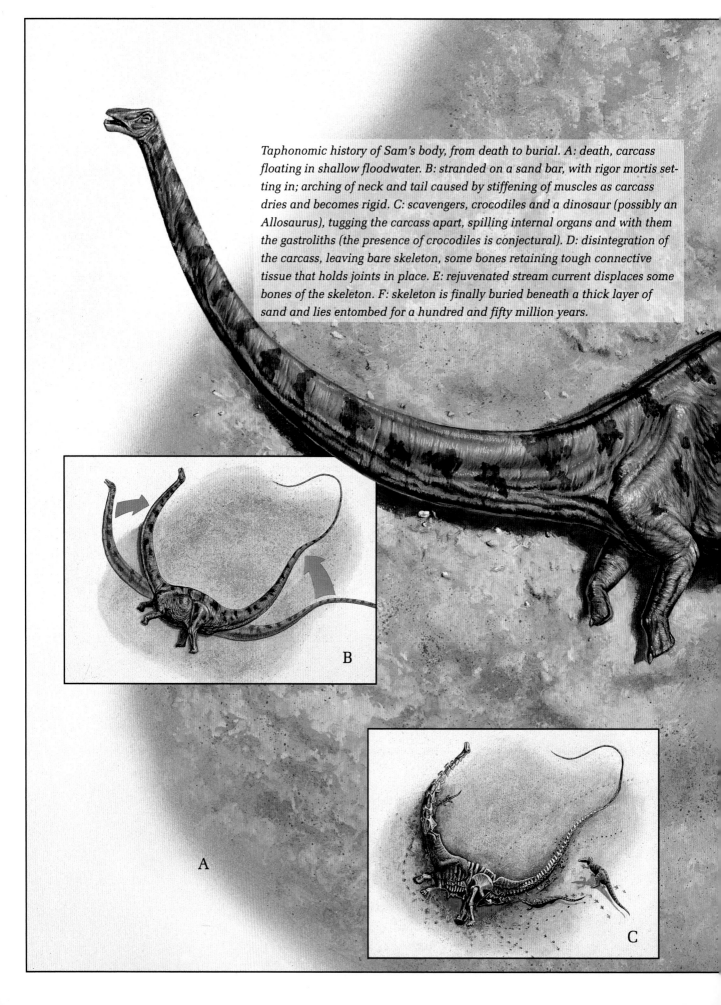

Taphonomic history of Sam's body, from death to burial. A: death, carcass floating in shallow floodwater. B: stranded on a sand bar, with rigor mortis setting in; arching of neck and tail caused by stiffening of muscles as carcass dries and becomes rigid. C: scavengers, crocodiles and a dinosaur (possibly an Allosaurus), tugging the carcass apart, spilling internal organs and with them the gastroliths (the presence of crocodiles is conjectural). D: disintegration of the carcass, leaving bare skeleton, some bones retaining tough connective tissue that holds joints in place. E: rejuvenated stream current displaces some bones of the skeleton. F: skeleton is finally buried beneath a thick layer of sand and lies entombed for a hundred and fifty million years.

A

B

C

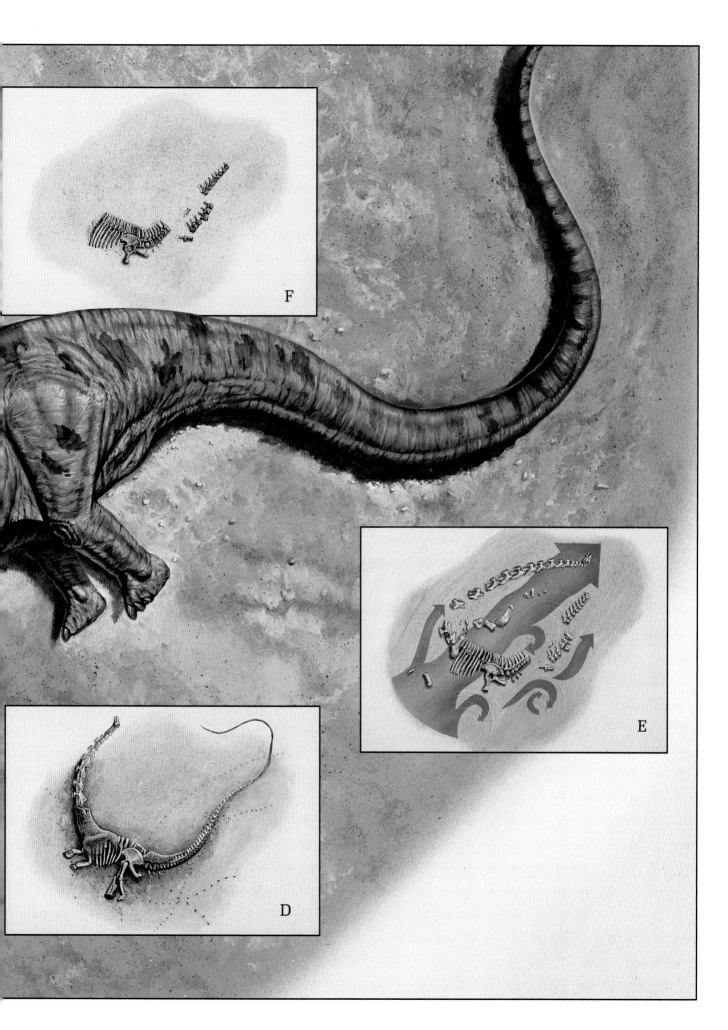

F

E

D

in the fossil record can we identify exactly how or why an animal died. But we can reconstruct the history of the body from the time of death until it was buried.

The site of death could not have been far away. Sam may have died right where we found the skeleton. Alternatively, perhaps Sam died on the banks of a river with a fast current that picked up the carcass and carried it to our site, where it became stranded on the sand bar. This explanation might account for the missing legs, which for some inexplicable reason became separated from the remainder of the carcass before it was buried. We have no useful information concerning the fate of the leg bones, making this idea attractive even if it's not easily supported.

The current must have subsided abruptly, leaving the carcass stranded above water level on the sand bank, oriented so that the tail pointed downstream, its torso and neck taking the brunt of the current. Before deterioration of the carcass had gone far, it lay exposed to hot sun and dry air. The muscles arched the back into a rigor mortis curve as the skeleton lay on its right side on the sand bar. Tendons and muscles were still on the skeleton at this point, and the gastroliths were still in the belly.

Soon, perhaps within a day or two, one or more large predatory dinosaurs, such as Allosaurus, discovered the carcass and scavenged a big meal. Here was a windfall for a carnivore, enough meat to last for days or weeks. As the scavenging continued, Sam's stomach bloated, spilling its contents, including the gastroliths, onto the sand bar. The scavengers loosened the joints in their quest for meat, especially the joints of smaller bones and the upper ribs from the left side of the rib cage, now dangling in the air.

Upper surfaces of the bones were stripped of their flesh, and they lay exposed to the hot summer sun. The bones dried and began to shrink. Their surfaces bleached and brittle, the bones began to crack and flake in the heat.

A storm rejuvenated the meandering, slow-moving stream, which was flowing east toward modern-day Texas. As the water level rose, sand began to pack in and around the slowly submerging skeleton. The bones still had dried muscle and tendon, holding the joints together. Ripples and sheets of sand covered the gastroliths at base level and surrounded the bottoms of the bones. The partially decomposed carcass blocked the stream flow; eddies began to swirl around the edges of the skeleton as water was diverted by this unusual barrier. Steep cross-bedding developed on the lee side of the skeleton, coalescing with other cross-bedded layers in this persistent sandbar.

The turbulent stream undercut some of the bones, especially the tail. Sections that had lost their connecting muscle settled in the sand, gently slipping away from the main line of the tail. Other parts of the tail, still held together by tendons, remained in place on their sides, in the rigor mortis curve. Leg bones, if they were attached at the time the carcass came to rest at this point, were undercut by the current action, too, and they were rolled downstream, away from the vertebrae and ribs. The neck became detached at the same time, perhaps sliding downstream a few yards until sand buried the long line of slender neck bones, maybe with the skull still attached. One vertebra near the base of the neck fell forward and downward on its face, onto a pile of gastroliths that had spilled out of the belly earlier.

Layer upon layer of sand piled around and on the bones, now under water. Where the bones were cracked and flaked, sand worked into crevices. Loose bones, such as the chevron and rib, tumbled end for end until they came to rest on the gentle downstream slope of the sandbar. They, too, were covered by the accumulating sand. Soon the entire skeleton was entombed, buried until one hundred fifty million years later, when Arthur Loy saw the bones on the edge of the mesa in the Ojito Wilderness Study Area of New Mexico.

This is the best story of the death and burial of Sam that we can construct for now. But laboratory preparation and study of the bone blocks themselves has only just begun, so the story may change — or be restated later with greater confidence.

Although taphonomy as applied by modern paleontologists usually embraces the period only from death to burial, events long after burial are also part of the history of Sam's bones. And so other questions arise: what happened to the skeleton after it was buried? How, specifically, did it become fossilized? Is any of this bone original, or is it all "replaced"?

The answers are surprising.

Sam's demise. The carcass of an adult Seismosaurus, initially dried and dehydrated, now submerged by floodwaters in a summer storm. An opportunistic Allosaurus tears into the newly softened flesh. This scene is a plausible dramatization of Sam's preburial history. The crocodiles are conjectural.

Chapter **8**

*Mysteries
of Fossilization*

Lot's wife, disobeying a warning, turned to look back on Sodom as she fled the evil city; for her disobedience, she was turned into a pillar of salt. In a moment, this woman changed from a living, breathing, moving human being into an inanimate rock. Instant fossilization.

For more than a century, paleontologists have explained fossilization in a sleight-of-hand manner. The explanation has stood, despite its weakness, because few professionals have concerned themselves with how fossils came to be fossils. The conventional wisdom that emerged with this laxity has allowed us to ignore what we do not understand.

Today most textbooks in paleontology delve into the subject of preservation at least superficially. The processes involved in preservation seem to be better understood for fossil plants and invertebrates than for vertebrates. Because the chemistry differs among these groups of organisms (shells do not have the same composition as bone, for example), the processes of fossilization must follow different paths. The subject of preservation chemistry of fossil bone is woefully neglected in textbooks and specialized journals alike.

However, a few paleontologists (taphonomists) have addressed the issue of fossilization directly, especially its early stages. These connoisseurs of fossil forensics have dedicated considerable effort to understanding what happens to an organism from the time it dies until long after it is buried as a carcass or a skeleton. The period from the death of an organism until it is discovered as a fossil, thousands and millions of years after it lived, encompasses its *taphonomic history*. Taphonomists generally emphasize the postmortem, preburial history in their studies, and only superficially consider postburial history, which is remote and largely chemical.

Consider the early taphonomic history of Sam, as described in the previous chapter. This sequence of events culminated in burial, itself a rare event, since most carcasses disintegrate and return to the soil or they are incorporated into the local food chain. This preburial history may take a few days or weeks, from the moment of death till burial, or even many months and perhaps a year or more. This is the usual scope of interest to taphonomists.

The remainder of the fossilization process and subsequent exposure through erosion or excavation is the postburial history. If a mammoth carcass and skeleton laid on the ground for ten years before the bones were fully buried, and it is discovered 10,000 years later, the period of burial is 9,990/10,000—or 99.9 percent of its taphonomic history. If it took ten years for Sam's bones to be fully buried, and because Sam lived roughly 150,000,000 years ago, the postburial years constitute 99.999995 percent of the taphonomic history.

Probably most organisms that are successfully buried (avoiding full recycling into the ecosystem) never become fossilized. In-the-ground changes destroy most organic materials, leaving nothing but chemical traces, if anything, as evidence of the past existence of organic remains in the rock. Only rarely do organisms or parts of organisms fossilize, which is really just another way of saying their disintegration is interrupted. We know almost nothing about the postburial history that leads to fossilization. We depend on untested generalizations to explain how this process of petrification proceeds. Such generalizations usually assert that preservation of fossil bone is a process of "molecule-by-molecule replacement"—a convenient, but ultimately vacuous, explanation that originated in the scientific literature over a century ago. In truth, the changes that occur after burial are complicated and not so easily dismissed.

For bones at least, the process probably never involves molecule-by-molecule replacement. In a broad sense, these are really questions of preservation chemistry: How are the fossil bones of a dinosaur different today from what they were 150 million years ago? Are they replaced? What, after all, does "replacement" mean?

In the mineral sense replacement is "the development in an old mineral of a new one that differs from it wholly or partly in chemical composition" (*Funk and Wagnalls College Dictionary*). With respect to shells of mollusks, for example, this means that one crystal form of calcium carbonate ($CaCo_3$) may replace another crystal form of the same chemical composition: calcite may replace the original aragonite in the shell of a particular clam. Or the chemical composition itself might change: pyrite (FeS_2) might replace calcite ($CaCo_3$), although the superficial form of the original shell is maintained. Is that how bone becomes fossilized, or "petrified"? Is the process a change from bone to stone, as we imply when we use the term *replacement*?

Most conspicuous is the form or shape of a fossil. Its shape resembles a fern frond that is compressed; therefore it must be a fern fossil. Or, its shape resembles the form of a femur; therefore it must be a femur. Often the shape is definitive, and we need go no further in our examination of the object to determine its identity. The form, or shape, of a fossil is thus its most important property. This feature alone becomes the basis for identifying a fossil, the attribute we examine when we have to make a decision whether to keep it (deposit it in a repository) or discard it.

But the two dimensional and three-dimensional traits are sometimes obscure until we get a "feel" for what we're looking for, as is often the case in the discovery of dinosaur track sites. Shapes can also be misinterpreted easily, leading to embarrassing mistakes. Form alone may not be sufficient to prove an object is even a fossil—much less a particular kind of fossil. We're often shown pseudofossils, or "foolers": they might have the correct form (usually a long stretch of the imagination), but they lack texture and structure, and on full examination show inappropriate chemical composition or association with the wrong kind of rock. They have form but not substance. The process of fossilization thus must preserve more than form.

Sometimes paleontologists, like medical doctors, want to see the internal structure of a bone — the microscopic details of its anatomy. A medical doctor examines under magnification a tissue sample of bone (a biopsy), perhaps prepared with a special cutting instrument called a microtome. Or, if there is grave concern, the medical specialist might want to examine the structure of a bone (or any other tissue) under much higher magnification, using electron microscopes for resolution smaller than the wavelengths of light used by an optical microscope. At all levels of magnification, the doctor can analyze the internal structure of the bone; cells are evident, and at high magnification internal anatomy of the cells is discernible. Paleontologists can perform the same task, using samples cut not by a microtome but by precision saws that can cut a fossil bone slice to a thickness of twenty microns — so thin that light can be transmitted through it. This is called a "thin section" of a rock or fossil. Thin sections are used in all aspects of geology for microscopic examination of composition, texture, and geological history. They are as fundamental to geology as tissue sections are to medicine.

Thin sections of fossil bone, examined under a microscope at magnifications of 10× or 100×, reveal incredible detail: individual bone cells (more properly called lacunae), often as finely preserved as in a tissue section of modern bone. The boundaries of the bone cells are crisply defined, the edges as sharp as when the animal was alive. This phenomenal preservation may be indiscernible to the naked eye because of the infilling of secondary minerals into the pore spaces of a fossil bone. Such minerals grew there long after the animal died, occupying the spaces — ranging in size from the internal marrow cavities to the microscopic lacunae — where fats and fluids and other tissues were contained in life. Often (but not always) the pore-filling mineral is quartz (SiO_2), and that is what makes the fossil bone heavy, and therefore "petrified." The original cells are still intact, however — unchanged, *not replaced* — even though the pore spaces are full.

Besides the filling of void spaces, which makes the bone heavier (denser) than an ordinary dried-out bone from a modern animal, fossil bone differs in the way it breaks. "Green bone" breaks in curved fractures. Fossil bone, on the other hand, breaks in rectangular patterns, and the breaks feel brittle, like the break in a piece of hard candy (for example, a large candy cane). This property of fossil bone arises from the loss of protein in its structure. For the rigid portions of living bone are actually a composite of the crystalline mineral hydroxyapatite and several hundred kinds of proteins. Proteins, especially fibrous collagen, impart resiliency to bones, and allow them to bend and flex when stressed. Since these proteins decay after burial, fossil bones lose the resiliency of living bone. They break all too easily. The natural inclination of a novice to test a fossil for its breaking strength results easily in breaks, at the very point of stress where the tester tries to break it. The test, producing brittle, rectangular fractures, naturally leads to the intuitive conclusion that the bone is indeed fossilized, or "petrified."

Thus, fossil bone generally has two essential properties that make it seem different from fresh, unfossilized bone: it is heavier because of the minerals that fill the spaces previously occupied by organic fluids, and it is brittle because the collagen proteins are destroyed. Most fossil bone also loses the creamy-white or buff color typical of living bone, changing usually to a darker color and often very dark shades of brown or black. The color changes (owing to in-filling of dark minerals) may be quite variable, even within a particular excavation site.

Dinosaur bones are almost always dark, usually close to black. Sam's bones, however, are close to white. Shades of light gray, light yellow, and light brown represent the range of Sam's bones recovered at the excavation site. Sam's skeleton, and several others in the Ojito Wilderness Study Area, are the only light-colored dinosaur bones from the widespread Morrison Formation I have ever heard of or seen.

Until now, I have simply described fossil bones, particularly dinosaur bones, without critically analyzing the process by which they change from living bone to fossil bone. The idea that fossil bone is the product of "replacement" always puzzled me, for one reason: the superb quality and detail of preservation. Fundamentally, there are no qualitative differences between the solid components of modern bone (say, a chicken bone left over from a picnic) and what is now found in those exact same places in fossil bone. Only the pore-filling minerals make the two seem different. If replacement is the process that produces fossil bone from living bone, then all anatomical details at the microscopic level should be lost or blurred. Indeed, the only feature remaining unchanged should be the gross overall anatomy.

Why do I make that argument? Because in mineralogical replacement, for example, of the mineral hydroxyapatite by quartz, one crystal substitutes for another (or several). Because of their crystal growth (which mineralogists call the crystal habit), no two minerals in crystal form occupy the same three-dimensional space. For two minerals as different as hydroxyapatite (the principal phosphate mineral in bone) and quartz (the most common secondary mineral found in fossil bone), the spatial relationships are markedly different. Now, at the level of examination in hand specimens, this difference should not matter: the surface texture and form of the fossil bone should remain essentially unchanged, even if the bone has been entirely replaced and is all quartz (at this extreme, the bone would appear glassy and smooth, a rare condition).

However, at microscopic levels of examination, if replacement is responsible for preservation of fossil bone, the differences in three-dimensional properties of the crystals should materially affect the quality of detail. The cells (technically the lacunae) should be destroyed, the evidence of their past existence obliterated or substantially altered. In essence, this destruction of detail at the cellular level would be an obliteration of the sharp boundaries between lacunae, and the boundaries between lacunae and pore spaces. Yet all fossil bone I have ever examined under a microscope shows no such loss of detail at the cellular level. The microscopic anatomy of fossil bone is spectacularly and crisply preserved. Parallels between modern bone and fossil bone are astonishing, even for dinosaur bones 225 million years old. Only an expert can distinguish fossil bone from modern bone at the microscopic level, and then the only clear distinction is color.

Owing to a seminar I gave at Los Alamos National Laboratory (described in chapter 3), chemists took up the challenge. George Matlack, Dale Spall, and Roland Hagan from the Lab, and Hilde Schwartz, my colleague from Dixon, New Mexico, launched an attack on the problem of preservation chemistry.

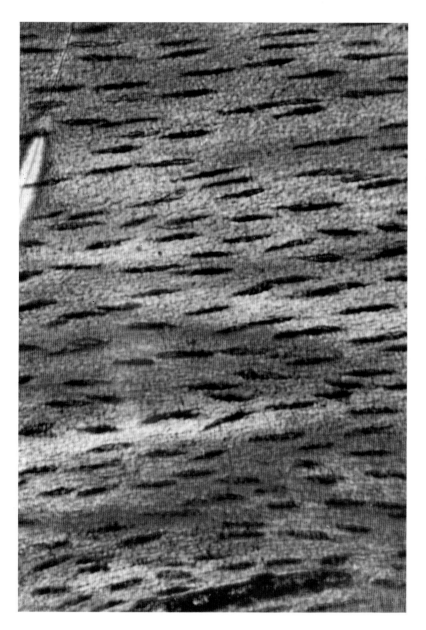

Thin section of leg bone from Coelophysis. This bone from a small carnivorous dinosaur, excavated at Ghost Ranch, New Mexico, is roughly 220 million years old (Chinle Formation; Triassic). The thin section was cut parallel to the haversian canals, which are now filled with clay minerals. The canals appear dark in this photograph, which was taken under polarized light. The brightly colored material is bone, mainly the mineral hydroxyapatite, largely unaltered.

Actually, this inquiry began somewhat later when Roland received a chemical analysis of a fragment of Sam's bones from Nate Bower, a chemist at Colorado College.

Nate's results were puzzling: for a series of major elements, the concentration in fossil bone almost perfectly matched the concentration in a sample of modern bone. Only silicon (in quartz, SiO_2) was different: high in the fossil and low in the modern bone. The similarities were astonishing. Maybe Sam's bones had not been petrified, someone suggested. No, I countered, these bones aren't any different, except for color, from thousands of fossil bones I had seen in my twenty years as a paleontologist. Sam's bones are not unique, I asserted; they instead represent typical dinosaur preservation.

But test results cannot be ignored. Testing is, after all, what separates science from speculation. If replacement were to account for fossilization, we reasoned, then the elemental comparisons between living and fossil bone should be substantially different. Nate's analysis planted the first real seeds of doubt.

Dave Mann, a Los Alamos specialist in making thin sections of rock samples (he had been one of the first to make thin sections of moon rocks), prepared thin sections of scraps of Sam's bone for us to examine. Like other dinosaur bones, the microscopic details were sharp: cellular structure was crisp, boundaries were intact, and the pore spaces were filled largely with crystalline quartz as in typical fossil bone. The high concentration of silicon in Nate's analysis was consistent with what we saw under the microscope, but how to explain the other elements? From two disciplines, analytical chemistry and mineralogy, we had complementary information. But from neither of these preliminary studies could we identify the mineralogical composition of the bone — or what it was that had the *form* of bone, as distinguished from the quartz infillings. Was the correspondence only a coincidence? Or, was the fossil bone real, its form *and substance* unchanged from its original composition?

To produce fresh samples of bone at the site, Dave built an ingenious, simple coring device from a drill press and spare parts in his laboratory. Lubricated by distilled water and powered by a generator at the excavation site, the coring apparatus cut into the rock like a tubular cookie cutter. Dave and several assistants drilled three solid cores, each about two inches in diameter, from one of Sam's bones that had been exposed but not treated with preservatives or hardeners. We had only exposed a small part of the bone, awaiting sufficient funding to commence a large-scale excavation, and we weren't positive which bone in the body it would be. Later we identified it as the seventh vertebra of the tail as counted from the hips rearward. At the time of drilling we did not know how thick this exposed bone would be (I had correctly guessed it was a caudal vertebra) or whether we would drill through a thick and dense section of

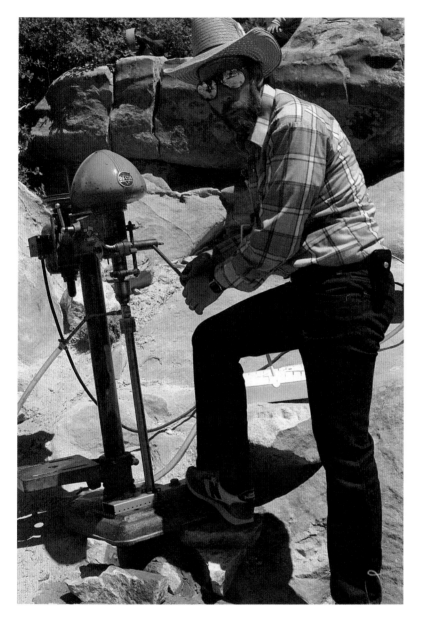

Dave Mann and his on-site coring device. The device is a modified drill press. The hollow coring bit was lubricated by distilled water to avoid contamination.

the bone. However, we decided to take the chance that we would get good sections in the cores, rather than expose the bone further and risk altering its in situ chemistry. The distilled water was for ensuring purity in the sampling: we did not want to introduce new chemicals to the bone and surrounding rock that might contaminate the cores and give false readings in our analyses.

These three cores became the samples for the first detailed chemical analyses of Sam's bones. Lab personnel treated the bones as though they were a precious metal. The cores were

Bone, in situ. This hint of bone was ex-posed during coring. The whole piece was later excavated and identified as caudal vertebra no. 7.

locked up in safes, paperwork was attached to every piece as though tracking nuclear material, and everyone handled the bone as though it were a newborn baby. In one meeting at the Lab, at least two dozen personnel showed up for the unwrapping of the first cores—an event akin to unveiling an artist's latest work.

The three original cores were all about a foot long. As they were crumbly, they were retained in foil jackets. Unfortunately, they contained only thin pieces of bone separated by sand that was only partially cemented. Clearly, we had not taken our samples from the main body of the vertebra, but from the edges where thin laminae support the external struts for muscle attachment. We confirmed this conclusion later, with excavation of the bone. In the cores, the thickest bone was less than a half inch—sufficient for analysis, but we had hoped for more. We kept the sand and sandstone intact so that we could analyze it too for its mineralogy and chemistry, as it was probable that the bone and its surrounding matrix were in a condition of dynamic equilibrium. Some of the chemical content of the bone would thus be leached into the sand, and some of the chemical content of the sand leached into the bone.

The dynamic equilibrium would be a condition of chemical adjustment corresponding to the temperature, pressure, and groundwater chemistry present near the time of exposure. Earlier in the burial history when Sam's bones were under considerable pressure beneath hundreds and perhaps well over a thousand feet of sediment and even deeply submerged beneath the waters of the Cretaceous sea, temperature and chemical conditions would have varied. The mineralogical and chemical properties of the dynamic equilibrium of bone and rock matrix would thus have varied too. Indeed, this is the guiding principal in understanding all chemical changes of rocks and minerals in the earth, and it applies equally to buried fossils: neither the surrounding rocks nor the fossils are inert. They are mutually affected by conditions of burial, their components interacting in chemical reactions that should be predictable. Assuming dynamic equilibrium as the guiding principal, the implied quest to identify the predictable chemical properties became our goal.

A less inquisitive group of scientists might have found Nate Bower's results sufficient to claim that the mineralogical part of Sam's bone (as opposed to the organic components such as proteins and fats) was essentially unchanged from its original chemistry. But we wanted to probe more deeply. For example, could the elemental chemistry be the same, but the minerals entirely different? And did the bone contain organic remains as well?

Understanding the chemistry of fossil bone is no easy problem. Living bone contains a multitude of chemicals in both organic form, such as fats and proteins, and inorganic forms such as the crystal structure of hydroxyapatite. In life, bone is continually growing and changing through the course of an individual's existence. Even after "growth" has ceased in adult humans, bone is alive, responding to stresses and changes just as other parts of the body change. Fossil bone preserves much of the structure or form of the original living bone, but what about its composition? Surely the organic portions on the whole are unstable, but some or parts of some may be durable enough to survive millions of years of burial. Amino acids have been recognized in invertebrate fossils from the Paleozoic Era. Is it possible that amino acids and proteins might be found in dinosaur bone, especially in bone so well preserved as Sam's? We faced two tasks: identify the mineralogical component of the fossil bone, and try to extract, even in small concentrations, the organic parts.

The mineralogical component of fossil bone is principally hydroxyapatite. It is a relatively simple crystal laid down by special bone cells responsible for growth and remodeling. This mineral contains the all-important calcium in combination with phosphate and carbonate. Its chemical formula is $Ca_5(PO_4,CO_3)_3(OH,Cl,F)_2$. This means that five atoms of calcium will be joined with three molecules of some combination of phosphate or carbonate and with two atoms of some combination of hydrogen-oxide, chlorine, and fluorine.

The proportion of fluorine is important. Fluorine may be incorporated into living bone in small concentrations, especially into the dentin of teeth, which is a specialized form of bone. The addition of fluorine strengthens bone, making it harder and denser, which imparts resistance to bacterial decay in teeth. With sufficient fluorine in its composition, the mineral name changes to fluorapatite, a variety of hydroxyapatite with essentially the same crystalline properties as its precursor. In isolation, a single crystal of hydroxyapatite or fluorapatite resembles a prism with six sides, expressed technically as a hexagonal crystal habit.

As bone grows, the mineral crystals grow initially in isolation, in a matrix of collagen fibers. The hydroxyapatite provides strength, the collagen proteins resilience and flexibility. Together, they make a remarkable material, capable of withstanding considerable compression, torsion, and stress, and forming the anchor for muscle action that propels the limbs and moves the various parts of the skeleton.

The highly organized microscopic architecture of bone develops according to direction of growth, stress fields in the bone, and position with respect to the internal anatomy of the bone. Much of the growth is under control of electrical stimuli generated from stress on and in the bone. The crystals of hydroxyapatite grow in an alignment parallel to the blood vessels contained within the bone, the haversian canals. The regular spacing and alignment of these crystals impart the strength of bone. Insufficient calcium in the diet impedes growth and maintenance of the hydroxyapatite crystals and may cause structural deformity, such rickets in children, because the bones cannot properly endure the weight of the body. They respond by bending and changing much as a pole will bend if overloaded.

In living bone, the orientation of the crystal axis of the hydroxyapatite is controlled by the direction of growth of the crystal. Generally, the orientation of the prismatic crystal is normal (that is, perpendicular) to the stress loading and parallel to the growth of the circulatory channels that carry blood and other fluids that bathe the internal tissues. Like the hydroxyapatite, the haversian canals grow normal to the stress loading; the bone that bounds the canals contains the regularly spaced hydroxyapatite crystals. Thus, the crystals and the orientation of the canals are parallel in living bone.

The canals are part of the form of the bone. They can be observed in microscopic detail in both living and fossil bone. They almost always appear to be unaltered in fossil bone, except their cavities may be filled with secondary crystalline minerals such as quartz. The hydroxyapatite crystals, on the other hand, are not part of the overall form of the bone, since they are exceedingly small and not readily visible even at high magnification under an optical microscope. According to expectations arising from study of crystalline materials in the geologic discipline called metamorphic petrology, the crystalline orientation of the hydroxyapatite should be altered with burial and compression in response to the new conditions; the major axis of orientation should be normal to the direction of compression during burial—which is likely to be very different from the direction of compression during growth. This is what we predicted for the hydroxyapatite in Sam's bones.

What about the collagen? The matrix in which hydroxyapatite crystals grow is a fibrous network of protein, mainly the variety of proteins collectively called collagen. This is the organic part of the structural components of bone. Collagens are proteins, complex molecules whose structure depends on chains of carbon in complicated linkages. These and other proteins are large organic molecules, thousands of times larger than the hydroxyapatite crystals (inorganic compounds with much simpler construction) which form in the protein matrix and give the bone its strength. Amino acids are simpler organic molecules that are collectively and systematically incorporated into proteins as building blocks. Some amino acids are stable over long intervals of geologic time, presumably preserved as remnants of the original protein. If amino acids could be recovered from invertebrate fossils of the Paleozoic Era, three or four times older than the dinosaur bones in the Ojito Wilderness Study Area, then we might be able to isolate amino acids and maybe even proteins or protein fragments from Sam's bones.

Because we needed more samples than the three we took on site with Dave Mann's coring rig, we decided later in the course of the experiments to take cores from one of the bones of the pelvis, the ischium. The ischium is one of the densest bones in a dinosaur's body (that is, it is less riddled with haversian channels and the bone is essentially solid), and it would provide excellent samples for analysis. We had excavated the ischium without application of hardeners or chemicals, thereby ensuring that we had not contaminated the bone unnecessarily.

By 1989 the multidisciplinary research projects on these bones were at a peak of intensity, and I traveled frequently to Los Alamos as a consultant to the Chemistry and Laser Sciences division. Many of the results led to one central conclusion: that we had original—truly original—bone, virtually unchanged ex-

Los Alamos National Laboratory scientists. Dale Spall, Dave Mann, and Lawrence Gurley (left to right in the foreground) take a solid core from Sam's ischium for chemical analysis.

Mysteries of Fossilization

cept for loss of some material (especially organic molecules) and only moderate modification of what was left (for example, the enrichment of fluorine in the hydroxyapatite). Los Alamos agreed to host a workshop for participants and anyone else with an interest in the subject of preservation chemistry. In March of 1989 we convened the Bone Chemistry Workshop, with three days of informal oral presentations by more than thirty participants, including scientists from beyond the confines of Los Alamos. The evidence from many angles was consistent: there was no "molecule-by-molecule replacement" in the fossils of Sam's bones.

Since then, some of these experiments have been completed, or otherwise abandoned. Others continue, especially in organic chemistry, as new techniques are applied for ever-more-complicated analyses. These may take several more years to finish. Many questions about the chemistry of preservation remain to be answered, particularly with respect to the timing and duration of each stage in the process. Moreover, the experiments were confined largely to Sam's bones. The burial history of Sam's skeleton is unique, the color is unusual, and many other variations and effects of burial are possible. We cannot, thus, globalize our conclusions with certainty — but we do suspect that other fossil bones are similar in the amount of original bone that remains.

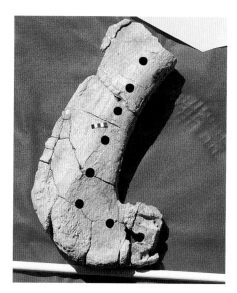

Ischium after samples extracted. The samples were extracted in the laboratory, after removal of the rock matrix. We chose the ischium for chemical analysis because its bone is among the densest in a dinosaur skeleton. Having fewer voids, it is subject to less in-filling by quartz during fossilization.

In the rest of this chapter I present an interpretation of some of the experiments, and then conclude with speculation on the process of fossilization itself as a sequence of events. The analyses focused on two major features of Sam's bones: the nature and orientation of the hydroxyapatite and associated minerals, and the quest to isolate proteins and other organic molecules.

The experiments on Sam's bones suggest that the hydroxyapatite is largely original. Most of the original hydroxyapatite is still preserved in the bone, albeit in slightly altered form, and most of the hydroxyapatite crystals retain their original orientation with respect to the haversian system of the bone. Two aspects of the hydroxyapatite crystals are, however, changed somewhat.

First, the chemical composition is enriched in fluorine (to 6 percent composition), changing the chemistry enough to prompt mineralogists to identify it as fluorapatite, a variant of hydroxyapatite that could as easily be called fluoro-hydroxyapatite. Six percent fluorine is much higher than that found in the bones of living vertebrates, and thus we suspect that dinosaurs did not, in the flesh, produce fluorine-rich bones. The addition of the fluorine does not change the size or shape of the crystals, but the fluorine seems to play an important role in the early stages of preservation (as described later).

Thin section of Sam's bone (normal light). The internal structure of the bone is sharply defined, even under the normal light used in this photograph. Photo courtesy Las Alamos National Laboratory.

Thin section of Sam's bone (polarized light). Polarized light enhances the definition of internal structure. The section was cut across the haversian canals; the dark centers of the elliptical figures are the former canals, now filled with clay minerals. The canals are surrounded by a dense growth of bone arranged in layers that grow parallel to the haversian system. The hydroxyapatite crystals and the shape of the bone cells that secreted them (too small to recognize at this magnification) are present in Sam's bones in a state almost identical to their condition at the time of death. Photo courtesy Las Alamos National Laboratory.

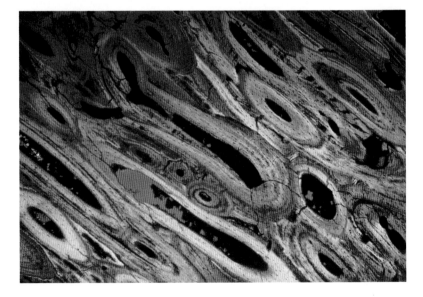

The other change in hydroxyapatite in Sam's bones involves the apparent growth of new crystalline material from the original hydroxyapatite; these new crystals grew after death from preexisting smaller seed crystals. Sam's bones now contain two sizes of hydroxyapatite crystals, the (presumably) smaller original crystals that are the majority of the composition and the secondary crystals that are less frequent but one hundred times larger. These larger crystals may have grown at the expense of other hydroxyapatite in the bone released to solution at pressure points in the fossil, or where fractures in the fossil allowed increased percolation of fluids that would dissolve and redistribute the elements at nucleation centers.

The analyses demonstrated two important surprises about the hydroxyapatite. First, compared with modern bone, the crystallinity is more pronounced, apparently because of the loss of protein, which in turn may have enhanced the secondary growth of the crystals after death (probably after burial). Second, the orientation of the hydroxyapatite crystals (the long axis of the crystal habit) still parallels the haversian system. This orientation should have changed with burial, pressure, and time.

The fact that the crystals remain unchanged in their orientation has three important implications: (1) the bone has probably not been buried deeply or subjected to high temperatures and pressures that would drive the reorientation; (2) the crystals are remarkably durable and not easily altered, a conclusion that should not be surprising in light of the remarkable strength and durability of living bone, one of the most durable materials in the natural world; and (3) the hydroxyapatite is mostly original (especially the smallest crystals) and unaltered even in their crystallographic orientation.

For me, this last implication was the clincher. It was proof that we have original composition, contradicting what some paleontologists have implied without direct evidence. The only material change is in the fluorine, which may have been the critical factor promoting preservation in the first place.

On the other hand, most of the organic composition is lost; only traces remain. But these traces theoretically can be isolated and identified. Quite unexpectedly, these remnants have been relatively stable through disintegration of the carcass, burial, temperature and pressure changes, and now exposure again to surface conditions. How they are preserved remains a puzzle, but the evidence for their existence seems convincing.

Bones contain more than four hundred different proteins, most of which are unique to bone and not found in other tissues. Collagen is the most abundant of the bone proteins, constituting more than half of all bone protein material. Collagen develops in the matrix between cells during bone growth. This protein fixes the position of nucleation of the hydroxyapatite crystals from specialized bone cells.

Los Alamos scientists Dale Spall, Lawrence Gurley, and colleagues isolated small quantities of protein from Sam's bone samples, which had been carefully handled to avoid contamination. The full analysis of the protein has not been completed, but one conclusion has been reached: the protein is not collagen. In any case, isolation of this protein, or proteins, from Sam's bones is the first instance of recognition of organic molecules from a dinosaur. Lawrence R. Gurley, Joseph G. Valdez, W. Dale Spall, Barbara F. Smith, and I published these results in 1991 in the *Journal of Protein Chemistry*.

The next logical step in the analysis is amino acid sequencing. It is the only means of positive identification of a particular protein. But it is a laborious and time-consuming process to establish the exact order of several thousands of amino acids; it is not clear whether this work will be done any time soon — if ever. However, the recent developments in genetic chemistry that use polymerase chain reaction enzymes for sequencing base pairs in DNA hold considerable promise for applications in paleontology.

Proteins may occur in greater concentration in other skeletons, preserved under different conditions of burial. Clearly, more work is in order on this general problem, if not particularly for Sam's bones. Proteins and genetic material have been isolated and analyzed from spectacularly preserved fossil leaves from Miocene lake beds (17 to 22 million years old) near Clarkia, Idaho. Paleontologist and popular science writer Stephen Jay Gould summarized in 1992 the occurrence of these leaves and their evolutionary implications. (See Gould's article for references to technical papers on the Clarkia fossils.)

With our discovery of noncollagen protein in a dinosaur bone comes many implications for future research on fossil bones. The potential applications are almost overwhelming. For example, we could begin to use comparative biochemistry of proteins and other biological macromolecules to test phylogenetic reconstructions of the various kinds of vertebrates — and thus determine who is related most closely to whom. We could better our understanding of the evolution of biological macromolecules, such as collagen and various bone proteins. We could learn to construct a geochemical history of a given site by calibrating the loss of sequentially susceptible organic compounds to pressure/temperature changes caused by burial.

And if there is protein, perhaps there is nucleic acid too. The futuristic science underlying the book and film *Jurassic Park* may not be all that far-fetched.

Out of these bits and pieces of evidence from Sam's bones, it is now possible to propose a rather tentative sequence of events leading to fossilization of bone in general. We can be sure of the beginning of the process and the final result: we start with living bone in a live animal, and we end with the fossil bone as it is discovered and removed from the ground. But what of all the events in between?

First, we must keep in mind that not all bones that initially become fossilized remain intact in the ground; surely, much or most bone that is initially preserved is lost to disintegration by groundwater under the destructive influence of high temperatures and high pressures of burial beneath hundreds and thousands of feet of sediment. Thus, the fossil bones we discover at the earth's surface may be only a small fraction of the bones that were preserved initially. The survivors had the right composition and nominally optimal conditions of burial; the sequential disintegration remained arrested or sufficiently slowed for a long enough period of burial to allow the bone to retain its original form and much of its composition until it reached the surface again as a consequence of erosion.

This history is precisely what we want to understand. At any point during burial, changes in the surrounding rocks may accelerate the disintegration and alter the fossil to such an extent that it is no longer recognizable: its form may be destroyed, and ultimately all chemical traces of its existence may be lost into and beyond the host rock.

First, the animal dies—the obvious starting point in our sequence of events. The carcass disintegrates and is subjected to the actions of scavengers. After an interval of time on the surface, which may range from only minutes to years, the carcass or, if it is stripped of all its flesh, the skeleton must be buried. Skeletons that remain on the surface (the usual and surely the majority of instances) fully disintegrate, their substance taken up by decomposers, scavengers, and soil-producing organisms. Only the skeletons that are buried and encased in a matrix of sediment or rock can become fossilized.

At some point in the process, soon after burial and probably within the first several years, percolating groundwater bathes the skeleton. Groundwater almost always contains fluorine in low concentrations. The bones take up the available fluorine by incorporation into the crystalline hydroxyapatite until they become saturated in this element. By this elemental substitution, the hydroxyapatite becomes fluorapatite. Indeed, this may be the single most important event leading to durability of the fossil bone. Without fluorine, perhaps all buried bones would be lost to further destruction. (We know that Pleistocene fossil bones between 5,000 and 1,600,000 years old also have high fluorine in their composition, but whether this pattern is universal has not yet been established.) The addition of the fluorine imparts resistance to further decay, similar to the resistance to bacterial decay that results from fluoridation of growing teeth. The bones become denser and harder.

Simultaneously, during the first several years after death, unstable organic molecules disintegrate. These includes fats and other lipids, amino acids, and proteins such as collagen. They probably disintegrate at predictable rates, each with different susceptibility to local conditions of burial. Most of these organic molecules are destroyed early in this process, but some apparently survive. Meanwhile, the loss of organic molecules and internal fluids leaves open spaces within the internal structure of the bone, mainly the conduits in the bone where the haversian system transmitted blood and other fluids. These voids are immediately available as tiny spaces for crystal growth.

At some point in the burial history, the voids are filled with new crystals generated from elements carried to the bone from groundwater seeping from the surrounding sediments or rocks. Typically, these crystals are quartz (SiO_2). Oxygen and silicon are, after all, the two most abundant elements in the earth's crust. Silicon is universally available in groundwater, and quartz easily precipitates under these conditions of burial. Eventually the pore spaces are filled with the new quartz crystals (or other minerals) and no space remains for growth or precipitation of additional materials.

This process of in-filling may be the second critical event (the first was the introduction of fluorine) making possible the long-term survival of the bone. Internally, the quartz crystals entirely occupy each pore space, preventing or at least retarding the penetration of additional groundwater and adding structural strength to the bones. With the addition of durable crystals filling the void spaces, the bones resist compression, an especially important attribute when the bones are in a compressible host rock such as clay or shale. If buried in a non-compressible rock such as sandstone or gravel, compression is less likely to change the bones, and their three-dimensional structure is preserved essentially unchanged.

These two events in the fossilization history (introduction of fluorine and filling of void spaces by mineral precipitation from groundwater) probably make the difference between preservation or destruction of buried bone. For even after burial, bone is subjected to a destructive environment. At least early in the burial history, decay organisms continue breaking down the bone by feeding on organic molecules and disaggregating the structural framework. Introduction of fluorine must thus occur early, too, and its effect would seem to be protection, perhaps by increasing the bone's density (or compactness) and inhibiting bacterial decay by making the internal structure inaccessible.

I suspect that the fluorine must be introduced almost simultaneously with burial, and that within a few tens or hundreds of years, during which time the bones may not be deeply buried and still remain subject to bacterial decay, the fluorine content is as high as it will get (approaching 6 percent in the hydroxyapatite, thus becoming fluorapatite). Also, I propose that bone not sufficiently invested with fluorine, for reasons of local variations in fluorine availability in the groundwater, will not fossilize; it is destined to decompose and its substance recycle into the sediments and solutions where no visible trace of its existence will remain. Thus, bones that get buried successfully will have their rate of decomposition arrested or slowed sufficiently to retard destruction. Other bones insufficiently enriched with fluorine, even in the same bone horizon, may not survive this early stage of decomposition, even though they are buried like the ones that survive.

Such differential loss of bone after burial, whereby some bones (or skeletons) survive post-burial decomposition and others are lost entirely, may seriously affect our efforts to census populations of extinct animals based on the abundance and density of their occurrence as fossils. In fact, disintegration of bone might explain why some sedimentary rocks contain no fossils. We often conclude that, because we find no fossil bones in a given formation or a given geographic area, fossil animals did not live there, or at least their skeletons were not buried there. This casual assumption may not be correct: in many cases, the differential loss of bone for want of fluorine may erase the evidence of their existence, leading us to incorrect conclusions. Paleontologist Robert Bakker has suggested a variation of this idea, elaborating on the notion of differential preservation of gastroliths and skeletons. By his argument, clusters of possible gastroliths in some geologic formations in the American West (especially the Morrison Formation) may have survived intact while the skeletons in which they were deposited and buried have dissolved and all vestiges of the bones have vanished.

Several observations lend support to these conclusions. First, in my work with Pleistocene mammals and reptiles, I have often found the bones to be soft and chalky. This is true in some sites (but not all) from Florida to Michigan, and from the East Coast to the deserts of the American West. This texture probably arises from disintegration of the bones during burial, and they would eventually be lost. Bones older than Pleistocene are almost always hard. Evidently, disintegration of buried bones is completed within the first 1,600,000 million years or so, and I suspect even within the first 100,000 years. In other Pleistocene sites bone is hard and well defined, even bone that is only a few thousand years old. A useful study would be to test for differential concentrations of fluorine in the chalky versus the hard Pleistocene bones. (In a different vein, it would be interesting to determine the effect of embalming chemicals on fluorine uptake in bone; it is possible that by embalming corpses we inhibit the incorporation of fluorine in buried bone, and consequently hasten the loss of skeletal material.)

Another observation that lends support to my ideas about long-term preservation of bone is that dinosaur bones are always filled with accessory minerals, such as quartz, that make them heavy; on the other hand, Pleistocene bones often retain their pore spaces, with little or no growth of accessory minerals. By this comparison, I suggest that the in-filling of void spaces in buried bone probably follows the introduction of fluorine but still occurs early, because all bone with filled void spaces retains most or all of its three-dimensional structure. If subjected to compression from burial, bone with extensive voids would compact or even become flattened, while bone with quartz filling the voids would resist deformation. The latter case seems to be the rule, although I have excavated some quartz-filled bones that were markedly compressed and deformed, but this is unusual.

Teeth seem to be the most durable skeletal material. Often we find only teeth and no bones; consequently, for some groups of vertebrates, teeth are the principal focus of attention in classification. This is especially true for mammals, and notably for Mesozoic mammals and mammal-like reptiles. Fortunately, mammal evolution at least partially followed a pattern of feeding specialization, which is reflected in the structure of the feeding apparatus, especially teeth. Thus, teeth are among the most desirable elements in a fossil site, and especially a site yielding mammals. I have always been puzzled about the differential preservation of teeth over bones. Teeth seem to be favored by the processes of fossilization to a greater extent than bone, perhaps because the internal structure of teeth (enamel especially, and dentin) is already dense at the time of death and burial and therefore resists decay more effectively than bone.

Finally, consider that fossil skeletons of adult animals are much more common than skeletons of babies and juveniles. But this surely does not reflect the actual population structure. Instead, it likely owes to the fact that the rapidly growing bones of young animals are more extensively invested with blood vessels and corresponding void spaces at death, subjecting these bones to more rapid disintegration before and after burial. Then, too, the bones of the youngest animals would likely be ingested with flesh by predators and scavengers, whereas the bones of adults might instead be picked.

These arguments (the chalky texture of some Pleistocene bones, the strengthening added by quartz infilling, a preservation bias in favor of teeth, and a preservation bias in favor of the denser bones of adult dinosaurs) all fall into the general category of "differential preservation." It may be the most difficult process to analyze because the differentiation results in an either/or product: either the buried bone is preserved (fossilized) or it is not. The null case cannot be studied, with the possible exception of the chalky Pleistocene bones that may represent fossils that would not make it beyond a few tens of thousands of years.

Bones that survive these early stages of disintegration after burial may still be subject to destruction and alteration from changes in subsurface conditions. The dinosaur bones buried in the Morrison Formation of the western United States were beneath hundreds and thousands of feet of additional sedimentary rocks during much of the Cretaceous Period, and many of these sediments were deposited beneath shallow and not-so-shallow seas. Introduction of marine waters to the underlying rocks certainly changed the conditions of fluid chemistry in the sediments surrounding the bones. The deep burial produced increased pressure, it elevated temperatures, and changed the nature of the fluids in the rocks surrounding the bones; all of these changes should have considerable effect on buried bones, although this speculation is hard to test. Perhaps the bones are already so resistant to destruction that these changes would have little effect, but I suspect that many bones are lost to these changes introduced from the processes of deep burial.

The elevation of temperatures through the geothermal gradient (an increase of ten degrees Centigrade per thousand feet of burial), probably affects organic molecules more than the inorganic composition of bone. Probably all vestiges of proteins and simpler organic compounds are destroyed by high temperatures; only bones that (like Sam's) were never deeply buried and therefore not subjected to markedly elevated temperatures will have residues of organic molecules. Relatively speaking, Sam's burial site has not suffered from great changes in pressure and temperature. Because the Ojito is in the thin edge of a deep basin (in which burial was deep in its middle) at the margin of the Colorado Plateau, its burial history is simple. We know from analysis of clay minerals that the temperature of these rocks never exceeded 100° C, a conclusion that indicates the rocks have never been subjected to heavy loading by deep burial. In other words, the sea was never very deep here, and there was never a great thickness of rocks overlying this site.

Until fossil bones are exposed at the surface of the earth, they are sealed in their host rocks in a state of quasi-equilibrium. With each physical and chemical change in the conditions of burial, the bones change too, but the hydroxyapatite seems to be highly resistant. Apparently, the destruction of hydroxyapatite is accomplished only with extreme changes, whereby the surrounding rocks are subjected to high pressures and temperatures that drive all chemicals to changes in crystal structure and mineral composition under new conditions of equilibrium.

Provided that the bones survive these gradual changes during their burial history, they may be freed eventually by erosion, as overlying rocks are stripped away by the forces of water, wind, and gravity. In the case of exposed sections of the Morrison Formation, the overlying rocks may once have been a thousand feet thick or more (depending on location). Stripped away by erosion stimulated by the uplift of the Colorado Plateau and surrounding areas, the eroded sediments were then carried away by streams and rivers.

The removal of overlying rocks brings new trials to buried

bone. Changes in ground water generated by rains and saturation zones in water tables may subject the fossil bones, now relatively shallow, to changes in acidity and alkalinity, and introduce elements that can precipitate as new minerals in and around the bone. Often we find fossil bones encased in extremely hard rinds of rock, which we collectively call concretionary matrix. This was the case for most of Sam's bones, but not all. I suspect that generation of concretionary materials around bones may occur relatively late in the burial history, not long before exposure at the surface of the earth. In dinosaur bones, this might happen during the last million years of their burial (remember, for dinosaur bones buried for 150 million years, like Sam's, the last million years of burial is less than 1 percent of its existence as buried fossil bone), and the process may not equally encase all bones in a given site or a given skeleton. For example, about a fourth of the surface area of Sam's bones was free of concretionary sandstone, and could be simply swept clean, or required only minor preparation to expose the bone. But most of Sam's bones were encased in the extremely hard concretionary layer, a rind that penetrated into the fabric of the bones and made laboratory preparation exceedingly difficult.

Eventually, bones that have been buried for eons may be exposed at the earth's surface. Overlying rocks stripped away by erosion, just to the level where the bones occur, the bones may remain in position and be only partially exhumed. This was the condition of the original eight tail bones of Sam's skeleton that we excavated in 1985. Part of the skeleton (the distal half of the tail) had been wholly exhumed, fragmented by erosion, and carried away as sediment, and part of the skeleton remained buried and unexposed; that was the part we subsequently excavated.

On reaching the surface, the bones are intensely affected by erosion and weathering, especially the action of freeze-thaw cycles that can reduce quality bone to fragments in a matter of months or years, especially bones that are not case hardened in a concretionary matrix. Once exposed, fossil bones do not last long; their usual fate is mechanical destruction, although some are collected by paleontologists and others may be recycled and become part of the newly deposited sediment in the stream beds that carry away the products of surface erosion. Forces of destruction are severe on exposed bones, a scientific tragedy in the sense that the bones have survived millions of years of burial only to be destroyed by the ordeal of surface exposure. Add to these natural causes the deliberate and accidental destruction of fossils by people, and it seems almost miraculous that any bone should survive long enough to become a museum specimen.

With so many situations working against preservation, we are truly lucky to find any fossil bones at all. Each stage in the preservation history is a kind of filter — or grim reaper. With each filter, surely no more than 10 percent persist. Probably 1 percent is a better estimate (but probably not as low as 0.1 percent). This latter figure would lead to an estimate of close to one trillion individuals in a given population of dinosaurs for a representative century.

Thus using the 10 percent figure, from a population of, say, a million individual dinosaurs that died in a given century during the late Jurassic, perhaps only 100,000 skeletons were buried successfully; 10,000 survived decomposition during early stages of burial; 1,000 survived deep burial; 100 survived shallow burial; and only 10 became exposed. Of those 10, only one survived on the surface long enough to be discovered by a paleontologist. That was Sam. One in a million.

Chapter 9

Sam's Heritage

Seismosaurus and the other spectacular dinosaurs of the Jurassic Period were the product of millions of years of evolution. Their reign came in the middle of the Age of Dinosaurs, which began in the Triassic 95 million years before Sam's time and lasted for another 85 million years, when dinosaurs finally became extinct at the end of the Cretaceous. During Sam's life, lasting perhaps a hundred years from hatching to that day 150 million years ago when Sam died of predation, choking, or some other cause, tectonic forces were barely perceptible in the middle of the North American continent. But the motion of the continents during the waning millennia of the Jurassic profoundly affected the world of the dinosaurs, nearly bringing the giant sauropods like Sam to extinction. Those events are almost lost to history, their traces scarcely recognizable and now remote from the desert of New Mexico.

In a broader sense, we now understand that the earth was restless and unstable during the Mesozoic. North America may have been relatively calm when Sam lived, but elsewhere radical changes were happening. Jurassic geography was profoundly different from the layout of continents and oceans today. Nor were the tectonic events at the end of the Jurassic the only time when dinosaurs and other life were affected by the continual restructuring of the earth's crust. Indeed, tectonic activity may be the ultimate driving force behind all evolutionary change, beginning even with the origin of life.

Dinosaur ancestors arose in the remote past of the Paleozoic Era, when North America and South America were joined with

Europe, Africa, and Asia in the giant supercontinent of Pangaea. Antarctica, Australia, and India were then still connected to Pangaea, too, with only minor hints that in the Mesozoic they would become islands. Pangaea, containing all the continental land mass of the Paleozoic world, lay on one side of the globe and the World Ocean on the other in a lopsided configuration that held few indications of changes to come. The subsequent restructuring of Pangaea in the Mesozoic Era was to drive a corresponding restructuring of life on earth, changing forever the nature of ecosystems on land and in the sea. In their own way, dinosaurs played a major role in the terrestrial ecosystem-in-flux.

The changes in life on land during the late Paleozoic and the Mesozoic do not read like a genealogy, one species following another in unbroken succession. That is what we should expect if our record of life on earth were complete, but alas, the book of life has many missing pages. Instead, the changes read from the fossil record are more like the evolution of music or culture: each generation builds on the successes and failures of past generations, recreating the heritage in its own way. Some of the changes seem to flow naturally, but occasionally the radical addition of a novelty wracks the seemingly complete and harmonious order.

Sam's heritage began long before the Jurassic Period, indeed, long before dinosaurs appeared. The origins of all dinosaurs and their ecosystems lay not in the Mesozoic Era, but more remotely in the waning stages of the Paleozoic Era, when much of Pangaea was dominated by reptiles whose ecosystems were complex and diverse. Reptiles prospered in the late Paleozoic, in the Carboniferous and Permian periods, and the ecosystems seem to have gradually become more complex through time. Changes in the faunas of the late Paleozoic led to improved success at dry-land existence, allowing reptiles to occupy deserts and uplands more successfully than before. These changes were important but not profound, for the animals and plants of the late Paleozoic had not been tested by major tectonic events like those that would follow in the Mesozoic.

Dinosaurs had not yet arisen, but their ancestors resided somewhere in a great profusion of reptilian species that dominated the closing epochs of the Paleozoic Era. The sail-back reptiles like Dimetrodon typify this time, but smaller and less remarkable animals dominated the terrestrial habitats. These were the mammal-like reptiles, such as Cynognathus, all living in a complicated food web that was surprisingly modern despite its antiquity (roughly 250 million years ago). Plant eaters were as large as pigs and cows, and highly specialized. Predatory reptiles were the primary and secondary consumers, feeding on the flesh of other reptiles. This complex ecological setting of the terrestrial vertebrates established an array of specialized niches in the Permian. The Permian vertebrate faunas of North America, Central Asia, and South Africa were nearly all reptilian, prospering in habitats that were relatively hostile to amphibians.

Even when all the land mass was locked into the supercontinent Pangaea, faunas found today in the fossil record of North America, Central Asia, and South Africa must have lived in regions separated by great distances. Yet they were remarkably similar. Tectonic forces had not yet created major barriers (like mountains and seaways) to dispersal. Permian reptiles flourished on land with unparalleled success. Their evolutionary history showed no sign of the pending disaster.

At the end of the Permian, and marking the end of the entire Paleozoic era, a disaster of untold proportions shook the biosphere. More than 90 percent of all species on earth went extinct. The niches occupied by the Permian reptiles were wracked by the sudden and catastrophic extinction that came closer to obliterating all life on earth than any other event since life became abundant in the Precambrian. The most profound change ever experienced by vertebrate animals was this calamitous extinction at the end of the Permian.

This Permo-Triassic extinction episode came without warning, a cataclysm that cleared many niches of their inhabitants but failed to destroy the potential of the biosphere to rebuild complexity. Abruptly the habitats became vacant on land and in the sea. The extinction was so abrupt and profound that this biological event itself marks not just the boundary between two geologic periods (Permian and Triassic) but the boundary between two geological eras: the Paleozoic Era and the Mesozoic Era. Newly evolved life forms that followed the extinction were drastically different, but the recovery was not immediate. Millions of years passed after the Great Extinction before the complexity was reestablished.

Following the Great Extinction and throughout the Triassic Period, tectonic convulsions shook the supercontinent. A rift developed in the southern part of Pangaea that eventually opened to form first the South Atlantic Ocean. Then the North Atlantic opened, like the jerks of a ragged and uneven zipper. Mountain-building generated by friction of the mobile land masses created barriers to atmospheric circulation and mixing of animal populations. Australia, India, and Antarctica separated from Pangaea and became island continents, their animals isolated like passengers in a raft.

Roughly 225 million years ago in the middle of the Triassic, after 20 million years of biological experimentation following the Great Extinction, dinosaurs first appeared. Their emergence had little immediate effect. Their exact ancestry is still a mystery, buried with the myriad of reptiles of the early Triassic called archosaurs. The early dinosaurs were small predators—about the size of a chicken. They were largely bipedal, nimble, and fleet of foot. They were, in turn, prey for the crocodile-like phytosaurs, giant floating and basking carnivores that probably overwhelmed their victims by ambush. The phytosaurs had teeth and jaws that must have inspired great feats of escape by the diminutive dinosaurs foraging or hunting along the shore and even venturing occasionally into water. Other archaic reptiles also competed for newly redefined niches in ecosystems based on the photosynthetic capacities of ferns, tree ferns, cycads, and conifers. Some coniferous forests of the Triassic grew to incredible heights. The profusion of giant logs at the Petrified Forest National Park in Arizona is evidence of one such forest.

The ecological setting for these early dinosaurs scarcely resembled the habitats that would follow. The earliest Triassic dinosaurs could easily have become extinct, snuffing out the dinosaur potential at an early stage, for their hold was fragile and tenuous. Their undramatic entrance onto the Mesozoic stage had little immediate effect on Triassic animals and plants. They did not sweep away their reptilian competitors' niches and replace them wholesale; instead the dinosaurs gradually specialized and diversified until by the end of the Triassic, their distribution was worldwide and their position secure — though still not conspicuous.

One specialized group of late Triassic dinosaurs became the ancestors of the giant sauropods of the Jurassic. These were the prosauropods, an aptly named variety of large plant-eating precursors to Apatosaurus, Diplodocus, Brachiosaurus, and their relatives including Seismosaurus. The prosauropods were giants of the Triassic and early Jurassic, with heavy builds and a general loss of bipedal locomotion, which returned them to the quadrupedal habits of their reptilian ancestors. Their relatively long necks and legs permitted them to feed higher in trees than the other herbivores, and they may have been able to balance on the hind legs to assume bipedal stance when feeding. The prosauropods were widespread and survived into the early part of the Jurassic, but their descendants, the sauropods, overshadowed them as the world of dinosaurs expanded and diversified. The prosauropod populations dwindled and eventually became extinct in the early Jurassic, their place in history more dramatic for their progeny than their contributions to their own contemporary landscape. From these gradual and modest beginnings, with roots in the prosauropods of the Triassic and early Jurassic, the sauropods emerged amid a bewildering array of other dinosaurs and reptiles, many gigantic.

At the end of the Triassic, huge geophysical changes once again occurred. The world of the dinosaurs went topsy-turvy in a reorganization that shook the terrestrial landscape to its foundations. Changes in the arrangement of the continents and oceans radically altered atmospheric circulation, the mixing of populations between regions or land masses and between the newly defined oceans and seas. These tectonic events disrupted and probably compressed the habitat gradients on the continents that had been stable for nearly a hundred million years.

As continents rattled and shook at the end of the Triassic, prominent chains of mountains took shape along the seams that once linked the Americas with Europe and Africa, while newly expanding oceans increased the isolation of the separate land masses. Seaways opened and closed in low latitudes, sometimes forming great expanses of oceanic waters. For example, the Tethys Sea persisted as a giant gulf that separated Europe from Africa, and North America from South America as the Atlantic Ocean continued to expand. Vestiges of the Tethys Sea today include the Caribbean and Mediterranean seas. Some of the Triassic land masses became isolated, their stranded populations of plants and animals confined to limited geographic areas. Such isolation could have promoted local episodes of speciation and variation among the land-dwelling inhabitants of the Triassic and early Jurassic. It might have prompted the beginnings of the sauropods out of prosauropod stock.

The sauropod dinosaurs probably originated from this period of intensified evolution, and they soon diversified to become the largest animals ever to live on land. They were present in Asia in the early Jurassic, but by the middle of Jurassic time the sauropods had become diverse and widespread, although not yet cosmopolitan. Their landscape in the middle and late Jurassic was radically different, more with respect to the animals with which they lived than the plants. The giant phytosaurs were gone, replaced by the earliest and equally deadly crocodiles. These aquatic predators seem to have secured their dominance in the shallow water habitats so well that dinosaurs never seriously challenged their occupation of the predatory niche of shorelines and open water. Flying reptiles (the pterosaurs) and the earliest birds fluttered or glided overhead, safe from marauding predators on the ground, successfully traversing open landscapes well beyond the reach of the meat-eaters.

The new niches were redefined and expanded. Dinosaurs capitalized on the additional complexity now possible with their single most important novelty: their size. Stegosaurs, iguanodonts, and ankylosaurs ambled along beside the sauropods, but their immensity (reaching several tons as adults) paled beside the adult sauropods of the Jurassic Period. Nevertheless, these herbivores exacted heavy tolls on their habitats, for their food requirements were immense. These were the usual prey for the giant carnivores of the Jurassic, typified by the awesome predator of the late Jurassic, Allosaurus. These giant meat-eaters in turn menaced at least baby and juvenile sauropods, if not the adults. And surely, just as in the ungulate herds of Africa today, sick and aged individuals from all populations of dinosaurs in the Mesozoic, including the sauropods, fell prey to these top carnivores.

The history of the sauropods, however, is perplexing. It seems out of synchrony with the history of other groups of dinosaurs. With dinosaur populations well established on all other continents by the middle of the Jurassic, the lack of sauropods in North America until near the close of the period almost defies explanation. But, from what we know of the fossil record, that was indeed the situation: sauropods were absent in North America until near the end of the Jurassic. When they appeared, however, the increase in their numbers and their ensuing diversification came astonishingly fast. Sauropod fossils appear first in North America at the bottom of the Morrison Formation. Everything we know about Jurassic sauropods in North America, the quintessential dinosaurs of the Mesozoic, comes from the Morrison Formation, the same body of rocks that yielded Seismosaurus.

Sauropods thus seem to have entered North America later than anywhere else. And despite their abundance in the Morrison Formation, only the broadest outline of their evolution can be discerned. All of the Jurassic species of sauropods were exceptionally large as adults, an order of magnitude larger than any other contemporary dinosaur. This difference marks an abrupt change in the ecological setting, for now the dominant plant-eaters far outstripped the largest predators in size. Relations between predators and prey surely became reorganized, but the nature of those changes is elusive. At least four families of sauropods entered North America from Eurasia by way of now lost land bridges in the late Jurassic: the Cetiosauridae (for example, Haplocanthosaurus), the Brachiosauridae (Brachiosaurus), the Camarasauridae (Camarasaurus), and the Diplodocidae (for example, Diplodocus, Apatosaurus, and Seismosaurus).

In North America the brachiosaurs (Brachiosaurus and Ultrasaurus) foundered, never reaching the abundance this family achieved in other parts of the globe. Brachiosaur skeletons occur only rarely in the Morrison Formation, but members of this family occur in large accumulations elsewhere, such as the Tendaguru beds of Tanzania in eastern Africa. Likewise, the

Six individuals in a herd of Seismosaurus.

cetiosaurs (Haplocanthosaurus) never became abundant in North America, where the more advanced sauropods seem to have eclipsed their possibility for success.

Camarasaurus, the most generalized sauropod, is the most abundant of the Morrison dinosaurs. More than any other dinosaur, Camarasaurus typifies the dinosaurs of the late Jurassic in North America. Its effect here was probably more extensive than that of any other dinosaur, for its populations probably outnumbered all other sauropods combined. This genus had several species, all closely related, and all built on the same basic body plan — heavy body, stout legs, relatively short tail and neck, and rather heavy skull reminiscent of a bulldog's head. Although successful in the late Jurassic of North America, the family did not proliferate into many genera and species in the fashion of the Diplodocidae.

The diplodocids, on the other hand, were diverse. Bones of these giants are common in the Morrison Formation, reflecting their abundance and diversity. Within this family belong the spectacular giants (and supergiants): Diplodocus, Apatosaurus, Barosaurus, Supersaurus, and Seismosaurus. They were the most highly specialized of the sauropods in North America, perhaps the most highly specialized sauropods ever. These dinosaurs had exceptionally long necks and tails, small (horse-size) and delicate skulls, and remarkably complicated adaptations for locomotion in their pelvis, sacrum, and vertebrae.

The great diversification of sauropods in North America that produced the giants and supergiants of the late Jurassic is problematic. Until recently, most geologists and paleontologists painted a relatively uniform picture of the Jurassic ecosystem: lowland habitats, largely humid and lush with vegetation. However, this assumption has been radically challenged during the past decade. Defying the traditional notion that sauropods had abundant food supplies and never wandered far from water is a new portrait that puts them in a patchy habitat. In contrast to earlier ideas about the Morrison Formation, western North America seems to have been ecologically diverse, ranging from wet lowland or riparian habitats to utterly barren desert. Vast tracts of virtually plantless terrain were dissected by perennial streams where vegetation grew in abundance. These narrow strips or belts of forests provided the sustenance for the plant-eating dinosaurs, especially the sauropods. They probably fed almost constantly when food was available, but they might have been forced to trek across barren land to find new sources.

Immense lakes, some of them alkaline and poisonous, covered large parts of the modern geographic region called the Colorado Plateau. These lakes and the open deserts were barriers to dispersal and migration, restricting the mobility of the giant dinosaurs, and perhaps also promoting geographic isolation and subsequent genetic isolation conducive to speciation. Thus the diversity of the landscape in turn promoted the diversification of the Jurassic giants, including the sauropods.

The large carnivorous dinosaurs of the late Jurassic were efficient meat-eating predators and scavengers that followed herds of sauropods, waiting for opportune moments — just as lions do today with herds of zebra and other grazers in Africa. Young sauropods and the sick or old were vulnerable to predators, but adults were relatively safe from those annoying meat-eaters in the same way that adult elephants in India or Africa have little to fear from carnivores. The most abundant carnivore was Allosaurus, which in many respects was more capable of taking down prey than its more famous (and more recent) relative, Tyrannosaurus. Allosaurus was more lightly built than *Tyrannosaurus rex* and was generally smaller, but some individuals may have approached the five-ton weight of a small Tyrannosaurus and reached lengths of more than fifty feet. Moreover, the teeth of Allosaurus were sharper, thinner, and better equipped for slicing flesh than were those of Tyrannosaurus. While Allosaurus and its contemporaries Ceratosaurus, Marshosaurus, and Stokesosaurus may not have been capable of taking large sauropods except in coordinated pack-hunting efforts (if indeed such social organization was possible), these menacing carnivores commanded the attention of all the giants of the late Jurassic.

Contemporary plant eaters of the late Jurassic were the plated dinosaur Stegosaurus, the small bipedal Camptosaurus, and the various nodosaurs and ankylosaurs — armored dinosaurs that foraged low to the ground and avoided the large predators probably through camouflage and the impressive protection afforded by their bony plates. These herbivores were so strikingly different from the sauropods that their habitats must have overlapped seldom, or never — as attested by the fact that they are almost never found in the same excavation sites as sauropods. The only likely times of competition would have been during searches for nesting ground, or when food was scarce, such as during a drought. Otherwise, these smaller (but nevertheless large by human standards) dinosaurs probably spent more time keeping out of harm's way in the moving landscape of legs and tails than they did in direct competition with the real giants of the time.

Pterosaurs flew above and may have interacted with the dinosaurs. These airborne reptiles were not yet large in the late Jurassic when Seismosaurus lived, but they may have been numerous in some places. However, their fossil remains are frustratingly rare in the Morrison Formation; whether that rarity reflects the lack of original abundance in the Jurassic of North America or represents a preservational bias is uncertain. Similarly, birds (Archaeopteryx) had originated by the late Jurassic, but none have been found with the Morrison dinosaurs. Their absence in the fossil record in North America probably reflects preservational bias. Only six specimens of Archaeopteryx have been found, all from Germany.

This paradise, the Jurassic zenith of the sauropods in which Sam lived, could not last forever. Changes in the forests of the late Jurassic, notably the advent of angiosperms (flowering plants), brought profound change to the world of the sauropods. And as the Atlantic Ocean expanded, South America, Antarctica, and Australia became more isolated, further restricting dispersal from one land mass to another.

For reasons still unexplained, the sauropods almost perished at the end of the Jurassic Period. Sauropod numbers and diversity diminished so dramatically that their existence for the next 80 million years during the Cretaceous Period was inconsequential. They were relics from a lost world. Consider: in North America alone, thousands of whole or partial skeletons of sauropods have been documented in the Morrison Formation of the late Jurassic; fewer than a dozen or so have been documented from the entire Cretaceous of North America.

What caused this Sauropod Crisis at the end of the Jurassic?

No one knows for sure. Indeed, few scientists have taken a serious interest in the dwindling of the sauropods, an event in their history that almost brought them to extinction. But there is one very plausible answer: the great sauropods may have been done in by flowers.

The first burst of color in the Mesozoic came with the advent of the angiosperms, the flowering plants. This event of the early Cretaceous was possibly the most important evolutionary invention since reptiles conquered the land with specially adapted eggs that could be laid out of water. This was, in essence, the sexual revolution of the Mesozoic, for the plants that employed flowers to propagate engaged other organisms in the reproductive stages of their life history.

Before the angiosperms, the ferns, cycads, and conifers had dominated the Mesozoic. Their reproductive structures produce a naked seed, with little stored nourishment, but in plenitude. Dispersal of the small seeds of conifers, like the seeds and spores of countless plants since invasion of dry land in the Paleozoic, was accomplished by little more than wind. But wind pollination and wind dispersal were no guarantee of reproductive success. Whatever the cause of the crisis on the Jurassic landscape, new plants with covered seeds, the angiosperms, emerged with increasing prominence. The key was reproductive success, probably promoted by the simultaneous diversification of pollinators, the advanced insects.

Plants with flowers could advertise free meals in the form of nectar or pollen, attracting pollinators that would incidentally improve the likelihood of fertilization of the developing eggs which lay deep within the inflorescence. Assured of successful pollination by the unwitting insects that carried pollen from one flower to another, the angiosperms invested heavily in the life history of the developing egg by adding copious supplies of stored energy, producing the covered seed ("angio sperm") that carried its own food supply sufficient to nourish germination and sustain the newly emerged plant for weeks.

The earliest angiosperms, some of which resembled the modern magnolias of the southeastern United States, collectively introduced a novelty into the plant world: the dependence on animals for fertilization, and to a lesser extent, for

(Following pages) Footprints of Brontopodus birdi *from Nashville, Arkansas. These footprints of a sauropod from the early Cretaceous document that sauropods survived the end-Jurassic extinction of many kinds of sauropods. The tracks were made by a large adult the size of Apatosaurus. The genus Brontopodus was named on the basis of these footprints alone, but they were probably made by the dinosaur whose bones (discovered elsewhere) have been named Pleurocoelous.*

dispersal. This innovation, in turn, promoted increased variation in the genetic vigor of the plants, for now plants could have their eggs fertilized by pollen from members of the same species from (relatively) distant populations. Specialization followed, with the flowering plants diversifying in a dramatic expansion of new families and genera.

This adaptive strategy prompted one of the most enduring revolutions in the terrestrial ecosystem, supplanting the conifers and their associated faunas while opening a myriad of new adaptive niches for animals and plants alike. Insects proliferated, and in turn became the major food supply for the Cretaceous mammals, which similarly diversified. Eventually the insects became important as food for birds and fishes as well, promoting the respective adaptive radiations of these vertebrates in the Cenozoic Era. The novelty of flowering plants stimulated this wholesale reorganization of life on land in an ever-expanding story of competition and specialization.

Conifer forests suffered, becoming isolated and restricted to geographic regions far from the premium landscapes that the flowering plants so fully exploited. (Remember that in the Cretaceous, worldwide climate was likely considerably warmer and moister than that of today. The cool-temperate, subarctic, and dry regions that support conifers in abundance today would have been rare in the Cretaceous.) By the end of the Cretaceous, conifers had lost their dominance, and the flowering plants had overwhelmed the terrestrial ecosystems. The dinosaurs that were closely tied to the conifers dwindled, too, suffering dramatic losses in abundance and diversity, while other dinosaurs seem to have capitalized on the complexity of the Cretaceous forests. Whereas those dinosaurs expanded in diversity and abundance with the diversification of the flowering plants, the sauropods nearly perished. They were left out of the new wave of experimentation among the dinosaurs, instead hanging on in archaic habitats that became increasingly isolated. The sauropods were overtaken by more progressive dinosaurs of the Cretaceous Period, their time of dominance long since passed.

Although the history of the sauropods cannot be directly and unequivocally associated with the demise of the conifers, the pattern is certainly persuasive: as the conifers dwindled, so did the sauropods — almost to extinction as early as the transition from the Jurassic to the Cretaceous. Although they survived in isolated populations around the world throughout the 80 million years of the Cretaceous Period, the sauropods were inconsequential and, except for their size, inconspicuous. They never truly recovered the losses they suffered at the Jurassic-Cretaceous transition. Their decline was so profound that they were absent from North America for at least half (40 million years) the Cretaceous Period.

The last sauropod genus in North America is Alamosaurus, an emigrant from South America whose remains are known only from Utah and New Mexico and only at the end of the Cretaceous Period, 65 million years ago. Several other sauropod genera managed to hold on till the end of the Cretaceous on other continents, notably South America and India, but these genera were from the same family (Titanosauridae) as Alamosaurus. The remains of these genera are nowhere abundant, and their populations were probably sparse.

The sauropods, thus, became extinct at the end of the Cretaceous 65 million years ago along with all the other dinosaurs and a myriad of other life forms — but they were on their way out long before. The end-Cretaceous extinction was not as severe or as profound as the Permo-Triassic event that set the stage for the dinosaurs. Their exit may have been dramatic, especially if they were ushered out by the impact of an extraterrestrial body. Or, it may have been relatively gradual, occurring over several million years. In either case, the demise of the sauropods at the end of the Cretaceous was the final whimper in a long and spectacular history.

In that long history, Seismosaurus may have been one of the (if not the) supreme examples of a dinosaur order that brought forth the biggest animals ever to walk on earth. The particular animal, Sam, which I was privileged to excavate, may or may not be the only example of this magnificent genus that science will ever come to know. But what we know of this individual tells us that Sam's kind was of spectacular dimensions.

Chapter 10

Was Seismosaurus the Longest?

Grade-school children are not the only ones who want to know who was fiercest or biggest or longest in the bestiary of dinosaurs. Paleontologists want to know that, too. I don't believe that Sam was the biggest (read, heaviest) dinosaur. And no sauropod was the fiercest. But *Seismosaurus hallorum* could have been the longest—or at least the longest thus far known to science.

For the first eight decades of the twentieth century, Diplodocus held the record for length (a composite mounted skeleton at the Carnegie Museum of Natural History measured about 87 feet). Known from several nearly complete skeletons, the skull and teeth of Diplodocus are much like that of Apatosaurus (formerly called Brontosaurus). Diplodocus is sometimes called the "double-beam dinosaur" for the unusual anatomy of the chevron bones of the tail. Like its relatives in the family Diplodocidae, the short front legs and decreasing size of the vertebrae from the hips forward imparted a downward slope from the hips, in contrast with the nearly equal limb length in Camarasaurus (family Camarasauridae) and the extraordinarily tall front legs of Brachiosaurus (family Brachiosauridae). With a weight of only 15 to 20 tons, Diplodocus was the most graceful and delicately built sauropod dinosaur. It has been found only in the western United States, Morrison Formation (late Jurassic).

The large, heavy body of Barosaurus rivals that of Apatosaurus for massiveness, but the extraordinary lengthening of the neck carried the long-neck experimentation to an extreme among North American sauropods. A skull has never been collected, nor have several other important parts of the skeleton. The recently mounted replica of a reconstructed Barosaurus skeleton at the American Museum of Natural History (New York) has generated considerable controversy because it is shown rearing high on its hind legs. Paleontologists are divided on the issue: some say the sauropods routinely reared back on the hind legs and reached upward with the long outstretched neck to feed high in the trees, perhaps taking foliage from the forest canopy. Others claim that this posture could be held only momentarily if at all and that individuals browsed with the neck and head held more horizontally.

Barosaurus may have surpassed the length of Diplodocus, perhaps reaching 100 feet or more, although whole skeletons have not been found to test that suggestion. More heavily built than Diplodocus, Barosaurus probably reached weights that rivalled Apatosaurus, to 30 tons and more. Barosaurus has been found in both North America (Morrison Formation) and the Tendagaru beds of Tanzania (late Jurassic).

Diplodocus, Apatosaurus, and the other giants all towered over the Mesozoic landscape. But with Brachiosaurus we enter the domain of the supergiants. Paleontologists recognized Brachiosaurus as the supreme sauropod for its incredible mass; however its length of roughly 80 feet fell short of the record held by Diplodocus. Although a recent find of one new genus of sauropod has proved larger, Brachiosaurus is today the most well known supergiant.

The "arm dinosaur," Brachiosaurus was named for its extraordinary front limbs. In this way it resembles a giraffe more than the usual high-hipped sauropod. Taller at the shoulders than the hips and with an enormous neck adapted for reaching upward, the center of gravity was farther forward than for the other dinosaurs. The relatively short tail contrasts with the long and slender tails of the Diplodocidae, accounting for its overall length of "only" 80 feet.

Calculations of mass for Brachiosaurus have been widely disparate. The techniques are based on artistic restoration by sculpting a three-dimensional model of great detail at a predetermined scale. Depending on the interpretation of how lean or fat the individual should be, mass calculations derived from such restorations range from 50 to 80 tons. This makes it about twice the weight of Apatosaurus, or about seven to eleven times that of a large elephant. Both the low and the high estimate are, in my view, reasonable; not only would adult weight vary with age and overall health, but a single individual probably varied enormously from season to season—perhaps behavioral seasons, such as a mating season or a migration, as much as climatic seasons.

The mounted skeleton of *Brachiosaurus brancai*, excavated in eastern Africa and now displayed in the Humboldt Museum in Berlin, reaches its long neck upward in a dramatic and spectacular display. It is the largest real skeleton on exhibit in the world. The Berlin mount spans 74 feet (22.5 meters) in length, and the head reaches 39 feet (12 meters) above ground level. Older restorations showed Brachiosaurus neck-high in water, presumably to support its prodigious weight, and sometimes even feeding beneath the water's surface. At 80 tons Brachiosaurus was immense, but there is no evidence to favor such aquatic interpretations. Instead Brachiosaurus, like its smaller relatives, lived on dry land and fed on trees and smaller vegetation. This supergiant may have entered water occasionally, but not as its preferred habitat.

With the possible exception of a recently discovered new sauropod, no other dinosaur stood as tall as Brachiosaurus. Other sauropods have a different build: generally low in the front quarters, with an exceptionally long and slender neck that was oriented in life more horizontally than vertically and a long slender tail. Comparisons of Brachiosaurus with other sauropods of greatly different shape and ancestry are therefore problematic. Nevertheless, Brachiosaurus is still the only supergiant known from reasonably complete skeletons. It remains the standard for comparison with other supergiants.

With the largest geographic range among the sauropod dinosaurs except Barosaurus, Brachiosaurus has been identified in eastern Africa (the Tendagaru beds of Tanzania), Portugal (of late Cretaceous age), and western North America (where incomplete bones and partial skeletons were excavated in the Morrison Formation of western Colorado in the early 1900s).

Still, after a century of excavation and discovery in the richly fossiliferous beds of the Morrison Formation, Brachiosaurus remains one of the rarest dinosaurs in North America. But Brachiosaurus is more than just a geographic puzzle. With a center of gravity far forward from the hips, its body plan differs from all other sauropods; it belongs in a family unto itself.

In 1985, the same year that I first saw the bones of Sam, Jim Jensen announced that he had discovered supergiant dinosaurs in the Dry Mesa quarry in western Colorado, not far from where Brachiosaurus had been found eighty years earlier. He had excavated these new bones from the same (Morrison) formation. As with Brachiosaurus, the bones Jensen discovered had been isolated before burial. With only a few exceptions, no two bones in that quarry have been found close enough together to conclude with certainty that they came from the same individual.

Jensen coined names, which he formally published in the same year, for his extraordinary giants: Supersaurus, Ultrasaurus, and Dystylosaurus. Supersaurus and Ultrasaurus gained notoriety overnight. Dystylosaurus was no less deserving of fame but it was burdened with a more difficult name (it means literally "double-strutted dinosaur," in reference to the supports, called laminae, for the neural spine). And so Dystylosaurus languished in the public eye.

Jensen applied these three names to three different bones from the Dry Mesa quarry, each a type specimen. He assumed the bones belonged not only to three different individuals, but to three different and hitherto unrecognized genera. Even in the field of dinosaur paleontology, this was a bold claim. Two of these bones could in fact belong to one individual. Moreover, at least one of the bones might belong to a dinosaur genus that had already been recognized: *Brachiosaurus.*

Jensen's formal description in 1985 of one of the impressive shoulder bones (a scapulacoracoid) from the Dry Mesa Quarry as the basis for the new genus *Supersaurus* has gained wide acceptance as representative of a dinosaur related to Diplodocus and Apatosaurus (family Diplodocidae), but with a heavy build. Other bones from Dry Mesa have been tentatively assigned to Supersaurus (a huge sacrum, a neck vertebra, several tail vertebrae, and several bones from the pelvis), but these were all isolated and could also belong to other diplodocids from the same site — where perhaps dozens of individuals in this family have been recovered. Technically, only the scapulacoracoid (called the type specimen) is known for this genus. Its anatomy clearly distinguishes it as a member of the diplodocid family, but the diagnostic characters that can be derived from the scapulacoracoid at the level of genus are dubious at best.

For decades, Diplodocus held the record for longest dinosaur, at 87 feet along the curvature of the spine from nose to tail. The giant scapulacoracoid from Dry Mesa that Jensen revealed in 1985 dwarfed the corresponding bones of Diplodocus and indicated much greater length, an astonishing discovery that captured the media worldwide. Its length may have exceeded 100 feet, perhaps 120 feet or more, making it a rival for the longest dinosaur; and its weight must have been at least as great as that of Brachiosaurus, 50 to 80 tons.

Jensen had to perform considerable restoration of the Ultrasaurus vertebra, but its proportions are dazzling, too. Except for larger size, it closely resembles the dorsal vertebra of Brachiosaurus. The vertebra has an undivided vertical spine, and its anatomy closely matches that of Brachiosaurus. Depending on which position in the vertebral series one compares it with, its osteology ranges from nearly identical to somewhat different. The similarity has led some paleontologists to declare Ultrasaurus a junior synonym of Brachiosaurus. Only an articulated skeleton, or at least a section of the vertebral column will resolve this uncertainty about Ultrasaurus. I prefer to recognize it as a valid but poorly known genus. Regardless, the individual animal was prodigious, surpassing the largest of known brachiosaurs.

Although these new names from the Dry Mesa quarry have become commonplace in dinosaur literature, we remain close to ignorant about the actual skeletons, their way of life, and their relationships to other dinosaurs. In my view, one bone does not make a beast. Ultrasaurus may, in fact, be a large species within the genus Brachiosaurus; Supersaurus may be closely related to Diplodocus.

Dystylosaurus, however, is a genuine puzzle. It seems to have no close relatives. The name was based on a huge dorsal vertebra from near the sacrum; its anatomy suggests affinity with the Diplodocidae. Certain features of the supporting struts indicate resemblance to the undescribed giant sacrum recently excavated from Dry Mesa and called Supersaurus in press releases. The most distinctive feature is the pair of support struts (as compared to only one in other sauropods) on each side of the neural spine, a condition that seems to be present on the "Supersaurus" sacrum. Regardless of its taxonomic status, this individual was a supergiant, ranking in size with Ultrasaurus, Supersaurus, and Brachiosaurus. Dystylosaurus was as large as Supersaurus; if my family assignment is correct, it too was exceedingly long, in keeping with the general proportions in the Diplodocidae.

Nonetheless, all three of these Dry Mesa genera could be designated *nomen dubium,* a taxonomic term signaling that insufficient information is available to adequately distinguish them from other taxa.

With their extraordinary size, the bones identified as Supersaurus and Ultrasaurus have naturally inspired considerable speculation, artist interpretations—and misconception—all stemming from the widespread belief that paleontologists can recreate an entire animal from a single bone. This deception, which we paleontologists, alas, willingly perpetuate as part of the mystery and romance of paleontology, delivers us into the realm of intuition more than objective fact.

Our ability to reconstruct an animal's skeleton is really a question of confidence. If we have the entire skeleton, then its reconstruction is not especially difficult; we must only assemble the bones and place them in reasonable orientations. If we have half the skeleton, then we have to fill in the missing parts with educated guesses. If we have only a leg, or only a foot, or only a single bone, our ability to rebuild the animal with accuracy is highly questionable. If we happen to have a diagnostic bone from a well-known species, then the problems of reconstruction are somewhat ameliorated. However, if the single bone, is the *only* bone known from that species, or genus, then reconstruction of a skeleton takes on a decidedly magical quality. Similarly, artistic rendering of restorations showing animals in life are equally suspicious when based on only a single bone.

In the case of Ultrasaurus, the single (dorsal) vertebra was the type specimen of the new genus. Vertebrae tend to show differences between genera, so this aspect of Jensen's work is not overly controversial. A giant scapulacoracoid found in the same quarry (which was not tied with certainty to the same individual because it was detached and isolated) did not bear sufficient anatomical detail to separate it from Brachiosaurus. But artists have combined the two bones, and thereby arrived at a putative full-body Ultrasaurus of immense proportions that dwarfs Brachiosaurus. The scapulacoracoid could just as likely belong to Brachiosaurus. Without better evidence that this bone belonged to Ultrasaurus, any skeletal reconstruction that incorporates it with the type specimen (the dorsal vertebra) is suspect.

Some artists have done the same for Supersaurus. Here too, they have combined a single scapulacoracoid (in this case, the type specimen) with tail vertebrae and other bones in the quarry and arrived at a composite that may not be genuine. Other artists have built the animal from the scapula alone.

To confuse matters even more, some paleontologists (and artists) have argued the merits of certain behavioral traits, such as the ability to rear up on the hind legs, from these reconstructions and restorations. At first glance, this seems like a reasonable approach. But, we do not have legs of Ultrasaurus, nor do we have sacrum or hips that might give clues to the anatomy of the legs and therefore the range of adaptations in locomotion. Without the legs or hips, arguments concerning posture have no merit whatsoever. Restorations *might be* correct, but without evidence, we can have little confidence in them. All restorations introduce considerable fantasy, even when based on complete skeletons; a restoration based on a single bone is, in my view, 99 percent fantasy and 1 percent science. In one case, for example, dimensions were calculated for limb bones of Supersaurus, based entirely on isometric scaling of the skeleton based on dimensions of the isolated scapulacoracoid, in comparison with other members of the family Diplodocidae. Such derived "data" are largely mythological; the margin of error in those dimensions is so great that the numbers cannot be trusted.

Naming and generic disputes aside, the bones excavated from Dry Mesa were indeed enormous, dwarfing all other sauropods except Brachiosaurus and Seismosaurus. Reliable calculations of the size of Ultrasaurus are almost impossible. But quite likely it was considerably longer than Brachiosaurus, perhaps well over 100 feet in length. Its mass was correspondingly large as well; compared to the mass of its closest relative, Brachiosaurus (50 to 80 tons), Ultrasaurus probably weighed 60 to 90 tons or more.

To date no one has attempted to restore Dystylosaurus, probably because of its difficult name. According to my reckoning, Dystylosaurus in restoration should resemble the diplodocids Apatosaurus, Diplodocus, and Barosaurus.

How does Seismosaurus compare with the supergiants of Dry Mesa? Our Ojito excavation produced far better data for understanding Seismosaurus than did the excavation at Dry Mesa for understanding Supersaurus, Ultrasaurus, and Dystylosaurus. In the case of Sam, the vertebral column from the base of the neck to the middle of the tail was in continuous articulation except

Comparison of the hip regions of Diplodocus, Apatosaurus, and Seismosaurus. (Drawn roughly to same scale.) Upper drawings show the hip regions from the top, lower drawings from the side. The open, nearly circular hole in each is the socket for the head of the femur. In all three genera, five vertebrae are fused in succession to form the sacrum; their lateral projections (technically, modified ribs) expand laterally to unite with the large upper bone of the hips, the ilium. The pubis projects downward and forward (to the left in this orientation), and the ischium downward and rearward (to the right). Note the tall vertical spines of Seismosaurus, and the distinct shape of the pelvic bones.

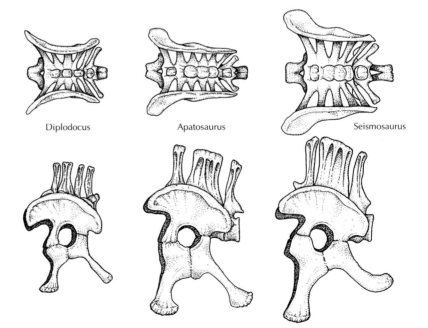

Diplodocus   Apatosaurus   Seismosaurus

Was Seismosaurus the Longest?

for an interruption in the tail. The skeleton also included several chevrons, a complete sacrum with all five vertebrae and five of the six bones of the pelvis (left ilium, right ilium, right pubis, left ischium, and right ischium) in articulation, three or four isolated neck bones, and some other isolated and unidentified bones. The dorsal vertebrae had ribs variously attached, some still in living position and others slightly displaced. In addition, found around Sam's skeleton were more than 240 documented stomach stones or gastroliths in the quarry, some in contact with ribs and vertebrae. This articulated, partial skeleton is the type specimen of *Seismosaurus hallorum*.

Because the bones were all still connected (articulated) or only slightly displaced, we can be sure we are dealing with only one individual (and only one species), without mixture of bones from different animals or different species. With only one individual dinosaur at the site, interpretations are decidedly less complicated than from the scattered bones of dozens of dinosaurs in the Dry Mesa quarry. Undistorted, uncrushed, and devoid of dark minerals filling pore space as in most fossil bones, the perfectly preserved three-dimensional nature of the Seismosaurus bones eliminates the need to restore them for study, further reducing the hazards of conjecture.

Beginning with the weekend excavation in 1985, the anatomy of the bones of the tail clearly distinguished this skeleton from Apatosaurus, Barosaurus, Brachiosaurus, Camarasaurus, and Haplocanthosaurus, leaving only Diplodocus, Ultrasaurus, and Supersaurus as possible candidates. The tall spines on the tail bones and their extraordinary length fit the diagnosis for the Family Diplodocidae, thus eliminating Ultrasaurus (Brachiosauridae) from consideration. That left only Supersaurus and Diplodocus, two members of Diplodocidae, as candidates.

Because no shoulder bones have been found at the Seismosaurus site we cannot make a direct comparison with the scapula of Supersaurus (the only bone confidently known for that genus). Among the hundreds of bones excavated from the Dry Mesa quarry—the type locality and only known site for Supersaurus—none of the tail bones resemble the ones from New Mexico. One set of tail vertebrae from Dry Mesa that Jim Jensen assigned tentatively to Supersaurus, probably correctly,

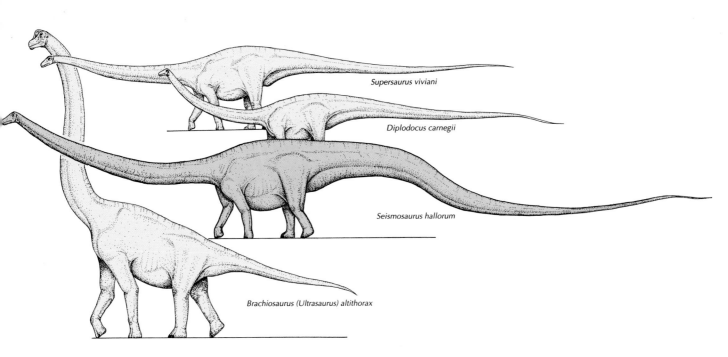

Supersaurus viviani

Diplodocus carnegii

Seismosaurus hallorum

Brachiosaurus (Ultrasaurus) altithorax

At 120–150 feet Seismosaurus was the longest dinosaur. Diplodocus, Supersaurus, Brachiosaurus, and its close relative Ultrasaurus approached lengths of 100 feet. The three supergiant sauropods—Seismosaurus, Brachiosaurus, and Supersaurus—weighed almost 100 tons (200,000 pounds), or the combined weight of fifteen large elephants. Diplodocus was smaller at 15–20 tons. Some paleontologists have argued that the weights of all four of these Jurassic giants were half these figures. In their estimation Seismosaurus, Brachiosaurus, and Supersaurus weighed 40–50 tons and Diplodocus 8-10 tons.

included eight in articulation from the middle of the tail of a large sauropod. They do not match the Seismosaurus tail bones, even allowing for variation, nor do any others from Dry Mesa. In addition, none of the other bones of the Seismosaurus skeleton resemble any of the bones from the Dry Mesa quarry. Overall, we can be sure that we are dealing with apples and oranges.

In life, the profile of Seismosaurus sloped gently downward in both directions from the tallest point at the hips. The huge bones of the hips and sacrum indicate an enormously heavy body, probably supported on legs no taller than those of Diplodocus. The extraordinary tail, perhaps 70 feet in length, had a peculiar kink about one-fifth of that distance from the hips, probably to lower the center of gravity. The general proportions suggest a neck of spectacular length, too, maybe as long as 65 or 70 feet. These figures, in my view, establish Seismosaurus as the longest sauropod dinosaur, with a projected length of at least 150 feet. Touching tail tips at the 50-yard line, two Seismosaurus adults could rest their heads on the cross bars of the goal posts at either end of a football field. (See Gillette 1991 for the full data and calculations.)

The excavated sacrum and hip bones of Seismosaurus, which had to bear the bulk of its mass during locomotion, hint at record weight as well. The larger the sacrum and pelvic bones, the heavier the individual. In at least one dimension, the 5-foot height of the sacral vertebrae (measured from the bottom edge of the vertebra to the tip of the spine), *Seismosaurus hallorum* exceeds the corresponding dimensions in all other sauropods. This dimension in Seismosaurus nearly doubles the height of the corresponding vertebra in Diplodocus and Apatosaurus among the dinosaurs in its own family (Diplodocidae), and probably exceeds the height of the giant (but crushed and distorted) sacrum from the Dry Mesa quarry identified as Supersaurus.

Because Sam's pelvic bones were enormous, I have concluded that Seismosaurus was heavy-bodied like Apatosaurus rather than slight like Diplodocus. Anatomically, Seismosaurus more closely resembles Diplodocus (for example, the concave ventral surface of the caudal vertebrae and the long, slender neural spines of the caudal and sacral vertebrae), but the proportions are closer to those of Apatosaurus. Thus, Seismosaurus was most closely related to Diplodocus, but because of its heavy proportions it outwardly resembled Apatosaurus.

Sam's vertebrae are all at least 20 percent longer than the corresponding vertebrae of Diplodocus (for which the figure of 87 feet is generally accepted). This translates into an estimate of overall length of at least 110 feet. That figure assumes direct proportionality with Diplodocus, or more technically, isometric proportions. Other dimensions of Sam's vertebrae and hips do not fit isometric proportionality. For example, the height dimensions of the caudal vertebrae and sacral vertebrae range to almost double the corresponding dimensions in Diplodocus and other members of the family. Quite surprisingly, the dimensions of the lower bones of the pelvis (ischium and pubis) are markedly disproportionate: anterior-posterior lengths almost double those of Diplodocus and Apatosaurus, but the heights are roughly the same. Thus, we cannot assume isometric proportions for Seismosaurus. Instead, Sam's dimensions are non-isometric, more technically *allometric*: growth of different parts of Sam's body proceeded at different rates of increase.

Sam was probably no taller than Diplodocus. This assertion should be no great surprise, since the taller a quadrupedal animal, the higher the center of gravity and the less secure its stability in locomotion. But the tall vertebrae and the massive pelvic bones indicate extraordinary length far beyond what we would conclude from isometric calculations. The neural spines supported the lifting muscles of the tail, with the sacrum as the anchor. Functionally the design resembles a suspension bridge, with the sacrum as the main pillar. The taller the neural spines, the longer the tail, and correspondingly, the longer the neck. Just how much longer is hard to determine, but certainly longer than the 20 percent longer calculated by assuming isometric proportions.

Similarly, the lower bones of the pelvis contributed to control of the muscles of locomotion, especially the lateral movements of the tail (the ischium) and of the abdomen (the pubis) which in turn provided the principal anchor for the neck. The enormity of the bones of the pelvis indicates neck and tail length markedly disproportionate. According to my calculations, Sam was between 128 and 170 feet in total length; 150 feet is a reasonable estimate. It is unlikely that the one fossil skeleton of Seismosaurus that has been discovered represents the biggest individual that ever lived. Other individuals of Seismosaurus were surely longer; thus a length of 170 feet may have been reached by some members of this genus.

Whether Supersaurus or other supergiants reached similar lengths cannot be established with the current state of knowledge of sauropods, but I regard that possibility as likely. At present, however, the better-documented Seismosaurus seems to be a serious challenger to Brachiosaurus for recognition as "the largest dinosaur."

Sam's mass is much more difficult than length to determine. Because of their weight-bearing function, legs and hips are the best indicators of mass in dinosaurs. Sam's legs are missing, so we must rely on the hips. Although the ilium and sacral vertebrae will take several more years to fully extricate from their surrounding rock, we have preliminary measurements that permit comparison with other dinosaurs. Sam's sacrum is at least as massive as the large sacrum attributed to Supersaurus at Dry Mesa. That sacrum and Sam's both exceed in all dimensions the sacrum of both Diplodocus and Apatosaurus, Sam's closest relatives. Similarly, Sam's sacrum is at least as large as the sacrum of

Brachiosaurus. Thus, Sam weighed as much as the other well-known supergiant sauropods, easily 100 tons and possibly more.

Overall, the questions Who was the longest? or Who was the biggest? may be fundamentally unanswerable for organisms that can be known only by their fossils. In a biological sense, Brachiosaurus and Seismosaurus, and nominally Ultrasaurus, Supersaurus, and Dystylosaurus, should all be regarded simply as giants or supergiants, or even as "the largest giants" or "the largest supergiants" because the largest individuals in all of these genera may never be discovered.

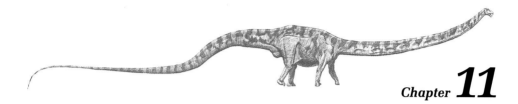

The excavation of *Seismosaurus hallorum* has ended, but the laboratory phase is just beginning. The four cervical vertebrae, the last bones to be discovered and removed from the quarry, are safely secured in the New Mexico Museum of Natural History in Albuquerque, where they will be repaired and prepared for study along with many of the other bones of the skeleton. Quite likely, no more bones will be found at the site and we will restore the quarry to a natural condition. Occasionally, paleontologists and hikers will visit the site, and perhaps eventually more bones will give a hint of their presence, owing to ongoing forces of erosion.

At least a dozen other dinosaur skeletons (or at least isolated bones) have been discovered in the Ojito Wilderness Study Area since the 1979 discovery of Sam. Some are sauropods. All deserve attention and some may warrant full excavation. One or several might be another individual of our new genus, *Seismosaurus*, but making such an identification may be impossible without considerable field investigation and excavation. Other skeletons could be new to science, too, representing new genera. Many of the secrets of the Ojito thus remain to be discovered.

Remote sensing tests, such as those we used at Sam's quarry, might be applied to advantage at these other sites. With each new field application, the technology will improve, and the limits of resolution will be pushed to smaller and smaller dimensions. Soon perhaps we will be able to confidently locate skeletons in the ground before turning the first shovel or bringing in the jackhammers. Such technology holds the promise of locating all the bones in a site, establishing the boundaries of an anticipated excavation, and even locating sites now buried and not evident from the ground's surface.

Meanwhile, we will turn our attention almost entirely to laboratory work on the bones, a task that will take roughly ten years for me and my (largely) volunteer crew to complete, absent direct funding of the work. Each bone must be freed from its stone matrix and prepared with painstaking precision before it can be fully studied. With the preparation of each new bone, more of Sam's anatomy can be described and compared to other dinosaurs, and we will improve our understanding of this one skeleton, this new genus. This slow and deliberate work requires skill and no small measure of patience. By the time this phase of the project is finished, we will have dedicated a major portion of our lives to this one remarkable dinosaur.

Simultaneously, experiments in the chemistry of preservation will continue to yield new and important data on the processes by which bone becomes fossil. At present, Los Alamos scientists are delving into the role of fluorine in the geochemical history of burial, to determine how fast bone can absorb this element and the limits of saturation under natural conditions. Our broad theory of fossilization will be tested, modified, and tested again.

Isolation of organic components such as proteins on Sam's bones (and, likely, on bones from other sites) will continue to occupy some of the researchers. The potential for expanding these applications into realms unimagined only a decade ago is high indeed. Proteins and genetic material, if they can be isolated and studied in detail, promise to be the first independent test of the patterns of evolution established on morphological grounds.

Meanwhile, we have a wealth of data on (and specimens of) documented gastroliths that could be compared with and used to test the identity of putative gastroliths at other dinosaur sites. All sorts of questions as to the selection and use by dinosaurs of these stomach stones remain to be answered. And with those answers will come a better understanding of dinosaur anatomy, physiology, diet, and possibly even behavior.

With each discovery and with each experiment, dinosaur bones give us a firmer picture of a remarkable chapter in biological evolution. Sam's existence as one of many sauropods in the Jurassic may have been minor and relatively insignificant in the ecology of the time, but the fossil remains of Sam are scientific treasures. What more may lay hidden in the sandstones of the Ojito? And what more may lay hidden elsewhere in the West?

Holiday centerpiece, December 1989. Wilson Bechtel appealing to the great god of dinosaur fossils for a complete skeleton, using the ischium and its core holes as a centerpiece.

References and Further Reading

This list of technical and popular articles includes references mentioned in the text, publications relevant to the Seismosaurus project, a few books and articles about dinosaurs in general, and one science fiction story (by Benford) inspired by this project. Articles and scientific papers that arose directly from research undertaken for the Seismosaurus project are marked with an asterisk.

Bakker, R. T. 1986. *The Dinosaur Heresies.* New York: Morrow.

Batory, R. D. and W. A. S. Sarjeant. 1989. Sussex Iguanodon footprints and the writing of *The Lost World.* In D. D. Gillette and M. G. Lockley, eds., *Dinosaur Tracks and Traces*, pp. 13–18. Cambridge: Cambridge University Press.

Behrensmeyer, A. K. and A. P. Hill, eds. 1980. *Fossils in the Making.* Chicago: University of Chicago Press.

*Benford, G. 1992. Shakers of the Earth. In B. Preiss and R. Silverberg, eds., *The Ultimate Dinosaur*, pp. 112–120. New York: Bantam. (Also published as Rumbling Earth, *Aboriginal Science Fiction*, Summer 1992, pp. 8–15.)

Bird, R. T. 1985. *Bones for Barnum Brown: Adventures of a Dinosaur Hunter*, with annotations by J. O. Farlow. Fort Worth: Texas Christian University Press.

Buffetaut, E. and V. Suteethorn. 1989. A sauropod skeleton associated with theropod teeth in the Upper Jurassic of Thailand: Remarks on the taphonomic and palaeoecological significance of such associations. *Palaeogeography, Palaeoclimatology, Palaeoecology* 73: 77–83.

*Chipera, S. J. and D. L. Bish. 1991. Application of X-ray diffraction crystallite size / strain analysis to Seismosaurus dinosaur bone. *Advances in X-Ray Analysis* 34: 473–482.

Colbert, E. H. 1962. The weights of dinosaurs. *American Museum Novitates* no. 2076.

——. 1968. *Men and Dinosaurs: The Search in Field and Laboratory.* New York: Dutton. (Reprinted in 1984 as *The Great Dinosaur Hunters and Their Discoveries.* New York: Dover.)

Czerkas, S. J. and E. C. Olson. 1987. *Dinosaurs Past and Present.* 2 vols. Natural History Museum of Los Angeles County.

*Dixon, P. R., W. D. Spall, and D. R. Janecky. 1991. Seismosaurus bone composition: Geochemical process indicator in a sandstone uranium province. Abstract with program. *Geological Society of America* 23: 17.

Dodson, P. 1990. Sauropod paleoecology. In D. B. Weishampel, P. Dodson, and H. Osmólska, eds., *The Dinosauria*, pp. 402–407. Berkeley: University of California Press.

Farlow, J. O. 1987. Speculations about the diet and digestive physiology of herbivorous dinosaurs. *Paleobiology* 13: 60–72.

Farlow, J. O., J. G. Pittman, and M. Hawthorne. 1989. *Brontopodus birdi*, Lower Cretaceous sauropod footprints from the U.S. Gulf coastal plain. In D. D. Gillette and M. G. Lockley, eds., *Dinosaur Tracks and Traces*, pp. 371–394. Cambridge: Cambridge University Press.

Galton, P. M. 1990. Basal sauropodomorpha—Prosauropoda. In D. B. Weishampel, P. Dodson, and H. Osmólska, eds., *The Dinosauria*, pp. 320–344. Berkeley: University of California Press.

*Gillette, D. D. 1986. A new place for old bones. *The Ghost Ranch Journal* 1(1): 2–6.

*——. 1986. A new giant sauropod from the Morrison Formation (Upper Jurassic) of New Mexico. Abstract in *Symposium on the Golden Age of Dinosaurs*, p. A18, North American Paleontological Convention IV, Boulder, Colorado.

*——. 1987. A giant sauropod from the Jackpile Sandstone member of the Morrison Formation (Upper Jurassic) of New Mexico. Abstract. *Journal of Vertebrate Paleontology* 7 (Supplement to 3): 16–17.

*——. 1990. Seismosaurus, the earth-shaker from New Mexico. *Ghost Ranch Journal* 5: 26–28.

*——. 1991. *Seismosaurus halli*, gen. et sp. nov., a new sauropod dinosaur from the Morrison Formation (Upper Jurassic/ Lower Cretaceous) of New Mexico, USA. *Journal of Vertebrate Paleontology* 11: 417–433.

*——. 1991. Paleoecological significance of gastroliths in a dinosaur excavation site. Abstract with program. *Geological Society of America* 23: 24.

*——. 1992. Ground-based remote sensing experiments at the *Seismosaurus* excavation, Brushy Basin Member, Morrison Formation, New Mexico. Abstract with program. *Geological Society of America* 24: 14.

*——. 1992. Form and function of gastroliths in sauropod dinosaurs. Abstract with program. *Geological Society of America* 24: 14.

*Gillette, D. D. and J. W. Bechtel. 1989. Pelvic and caudal anatomy of the giant sauropod dinosaur "Seismosaurus" (Morrison Formation, New Mexico). Abstract. *Journal of Vertebrate Paleontology* 9(3): 22A.

*Gillette, D. D., J. W. Bechtel, and P. Bechtel. 1990. Gastroliths of a sauropod dinosaur from New Mexico. Abstract. *Journal of Vertebrate Paleontology* 10(3) supplement: 24A.

*Gillette, D. D., J. L. Gillette, and D. A. Thomas. 1985. A diplodicine dinosaur in the Morrison Formation of New Mexico. Annual Symposium on Southwestern Geology and Paleontology, Museum of Northern Arizona, p. 4.

*Gillette, D. D., A. Witten, W. C. King, J. Sypniewski, J. W. Bechtel, and P. Bechtel. 1989. Geophysical diffraction tomography at the "Seismosaurus" site. Annual Symposium on Southwestern Geology and Paleontology, Museum of Northern Arizona, p. 10.

Gillette, D. D. and M. G. Lockley, eds. 1989. *Dinosaur Tracks and Traces.* Cambridge: Cambridge University Press.

*Gillette, J. L. 1994. *The Search for Seismosaurus, the World's Longest Dinosaur.* New York: Dial Press.

Gould, S. J. 1992. Magnolias from Moscow. *Natural History* 101: 10–18.

*Gurley, L. R., J. G. Valdez, W. D. Spall, B. F. Smith, and D. D. Gillette. 1991. Proteins in the fossilized bone of the dinosaur, Seismosaurus. *Journal of Protein Chemistry* 10: 75–90.

Harris, J. 1987. Introduction. In S. J. Czerkas and E. C. Olson, eds., *Dinosaurs Past and Present,* volume 1, pp. 1–6.

Jacobs, L. 1993. *Quest for the African Dinosaurs: Ancient Roots of the Modern World.* New York: Willard Books.

Janensch, W. 1929. Megenstein bei sauropoden der Tendaguru-schichten. *Palaeontographica* Supp. VII (1) teil 2, lief. 1: 137–143.

Jensen, J. A. 1985. Three new sauropod dinosaurs from the Upper Jurassic of Colorado. *Great Basin Naturalist* 45: 697–709.

——. 1985. Uncompahgre dinosaur fauna: A preliminary report. *Great Basin Naturalist* 45: 710–720.

——. 1987. New brachiosaur material from the Late Jurassic of Utah and Colorado. *Great Basin Naturalist* 47: 592–608.

*Johnston, R. G., K. Manley, and C. L. Lemanski. 1990. Characterizing gastrolith surface roughness with light scattering. *Optics Communications* 74: 279–283.

Lockley, M. G. 1991. *Tracking Dinosaurs*. Cambridge: Cambridge University Press.

*Manley, K. 1991. Two techniques for measuring surface polish as applied to gastroliths. *Ichnos* 1: 313–316.

*Manley, K. 1991. Gastrolith identification and sauropod dinosaur migration. Abstract with program. *Geological Society of America* 23: 45.

Martill, D. M. 1991. Bones as stones: The contribution of vertebrate remains to the lithologic record. In S. K. Donovan, ed., *The Processes of Fossilization*, pp. 270–292. New York: Columbia University Press.

McIntosh, J. S. 1990. Sauropoda. In D. B. Weishampel, P. Dodson, and H. Osmólska, eds., *The Dinosauria*, pp. 345–401. Berkeley: University of California Press.

Miller, W. E., J. L. Baer, K. L. Stadtman, and B. Britt. 1991. The Dry Mesa dinosaur quarry, Mesa County, Colorado. *Guide-*

book for *Dinosaur Quarries and Tracksites Tour, Western Colorado and Eastern Utah*, pp. 31–45. Grand Junction Geological Society, Colorado.

Paul, G. S. 1988. The brachiosaur giants of the Morrison and Tendaguru with a description of a new subgenus, *Giraffatitan*, and a comparison of the world's largest dinosaurs. *Hunteria* 2(3): 1–14.

Peters, R. H. 1983. *The Ecological Implications of Body Size*. New York: Cambridge University Press.

Schaedler, J. M., L. Krook, J. A. M. Wootton, B. Hover, B. Brodsky, M. D. Naresh, D. D. Gillette, D. B. Madsen, R. H. Horne, and R. R. Minor. 1992. Studies of collagen in bone and dentin matrix of a Columbian mammoth (late Pleistocene) of central Utah. *Matrix* 12: 297–307.

*Schwartz, H. L. and K. Manley. 1992. Geology and stratigraphy of the *Seismosaurus* locality, Sandoval County, New Mexico. *New Mexico Geology* 14: 25–30.

*Schwartz, H. L., K. Manley, and D. D. Gillette. 1989. Taphonomy and Paleoecology of the Upper Jurassic(?) "Seismosaurus" locality, San Ysidro, New Mexico. Abstract. *Journal of Vertebrate Paleontology* 9(3): 37A.

*Spall, W. D., D. R. Janecky, and P. R. Dixon. 1991. Seismosaurus bone composition: Proteins. Abstract with program. *Geological Society of America* 23: 96.

Spencer, F. 1990. *Piltdown: A Scientific Forgery*. London: Oxford University Press.

Stevens, C. E. 1988. *Comparative Physiology of the Vertebrate Digestive System*. New York: Cambridge University Press.

Stokes, W. L. 1942. Some field observations on the origin of the Morrison "gastroliths." *Science* 95(2453): 18–19.

——. 1987. Dinosaur gastroliths revisited. *Journal of Paleontology* 61: 1242–1246.

Weigelt, J. 1989. *Recent Vertebrate Carcasses and Their Paleobiological Implications.* Chicago: University of Chicago Press.

Weishampel, D. B., P. Dodson, and H. Osmólska, eds. 1990. *The Dinosauria.* Berkeley: University of California Press.

*Witten, A., D. D. Gillette, W. C. King, J. Sypniewski, J. W. Bechtel, and P. Bechtel. 1989. Dinobusting: Geophysical diffraction tomography in exploration paleontology. Abstract in program of Annual Meeting of American Geophysical Union.

*Witten, A., D. D. Gillette, J. Sypniewski, and W. C. King. 1992. Geophysical diffraction tomography at a dinosaur site. *Geophysics* 57: 187–195.

*Woldegabriel, G. and R. Hagan. 1990. Temporal and spatial relationships of diagenetic processes in the Upper Jurassic Morrison Formation, Colorado and New Mexico, and its implication to dinosaur fossils preservation. *Isochron/West* 55: 18–23.

Index

Designer: Heidi Haeuser

Text: New Baskerville

Compositor: Keystone Typesetting, Inc.

Printer: Oceanic Graphic Printing

Binder: Oceanic Graphic Printing